MOONG OVER MICROCHIPS

MOONG OVER MICROCHIPS

ADVENTURES OF A TECHIE-TURNED-FARMER

VENKAT IYER

PENGUIN
VIKING
An imprint of Penguin Random House

VIKING

USA | Canada | UK | Ireland | Australia
New Zealand | India | South Africa | China

Viking is part of the Penguin Random House group of companies
whose addresses can be found at global.penguinrandomhouse.com

Published by Penguin Random House India Pvt. Ltd
7th Floor, Infinity Tower C, DLF Cyber City,
Gurgaon 122 002, Haryana, India

First published in Viking by Penguin Random House India 2018

Copyright © Venkateshwaran Iyer 2018

Photographs by Venkateshwaran Iyer and Meena Menon

All rights reserved

10 9 8 7 6 5 4 3 2 1

The views and opinions expressed in this book are the author's own and the facts
are as reported by him which have been verified to the extent possible, and the
publishers are not in any way liable for the same. Names of some people have
been changed to protect their privacy.

ISBN 9780670090907

Typeset in Adobe Garamond Pro by Manipal Digital Systems, Manipal
Printed at Thomson Press India Ltd, New Delhi

This book is sold subject to the condition that it shall not, by way of trade
or otherwise, be lent, resold, hired out, or otherwise circulated without the
publisher's prior consent in any form of binding or cover other than that in
which it is published and without a similar condition including this condition
being imposed on the subsequent purchaser.

www.penguin.co.in

*To Meena, without whose support and
encouragement none of this would have been possible,
and to my dear friend Dilip Samel, who would have
enjoyed reading this book if he was alive*

With all its sham, drudgery, and broken dreams,
it is still a beautiful world.
Be cheerful.
Strive to be happy.

—Max Ehrmann

Man cannot discover new oceans unless he has the courage to lose sight of the shore.

—André Gide

There seem to be but three ways for a nation to acquire wealth. The first is by war, as the Romans did, in plundering their conquered neighbours. This is robbery. The second by commerce, which is generally cheating. The third by agriculture, the only honest way, wherein man receives a real increase of the seed thrown into the ground, in a kind of continual miracle, wrought by the hand of God in his favour, as a reward for his innocent life and his virtuous industry.

—Benjamin Franklin

Contents

Acknowledgements   xiii
Preface   xv

1. Where Am I Headed?   1
   To Leave or Not to Leave
   Changing Over

2. Search for an Alternative   14
   First Interest in Farming
   Working out the Finance
   A Cinematic Encounter and the Final Plunge

3. Searching for Land   27
   The Transition
   A Different Journey

Contents

4. Land at Last   36
   The Search Ends
   My New Address

5. Early Lesson in Farming   56
   The First Crop
   Selling Moong
   The Water Diviner

6. First Year at the Farm   74
   Groundnut Harvest
   First Anniversary

7. The Search for Rice   83
   The Scent of Rice
   System of Rice Intensification

8. The Present Scenario   91
   A Day at the Farm
   Recognition

9. Settling Down   107
   Breaking from the Trap
   Village Community
   Shh . . . The Gods Are Coming
   Spells and Curses
   The Mahalaxmi Temple Fair

Contents

10. Of Kerosene, Groundnuts and Subsidies   131
    Scams Here Too
    Twinkle, Twinkle
    Sarkari Troubles
    Murder
    Thefts

11. Doctor in the House   158
    The Death of Moru Dada

12. Snakes, Owls and Other Animals   169
    The Hissing Cobra
    The Dreaded Event
    Allahrakha
    Hen Log
    The Cats and Pepper

13. Village Economics and the Man Who Hates Banks   196
    The Man Who Hates Banks
    The Demon of Demonetization

14. Market Initiatives   209
    Lessons in Supply Chain
    The MOFCA Experience

15. Are You Happy?   224
    What's in the Future?

Glossary   237

Acknowledgements

I read somewhere that many events in our lives are shaped by close friends and relatives. There cannot be a more accurate way to describe the birth of this book.

What started as a series of emails to former colleagues in IBM soon transformed into the idea of writing a book to chronicle the events in my life—how I moved from the corporate world to the farming world and how I managed the transition from being a city dweller to a villager.

I got the jitters when my publisher asked me to write the acknowledgements for this book. How can I mention the long list of people who encouraged me to write my experiences and kept me going when so many publishers refused to bring out my book? What if I forget to mention someone and hurt their feelings? Yet, I shall try my best to remember all those who egged me on to finally finish it.

My friends from IBM—Sriram, Shrikant, Sanjeev, Radhesh, Nitish and Rajesh—who were supportive at each

Acknowledgements

stage and kept my morale up during the difficult stages of writing. My friends from outside IBM like Hema, Ujwalla and Anjali who kept encouraging me by giving examples of various writers who had managed to publish their works eventually.

This book would not have been possible without the support of Rao and Kishal from the USA who were among the first to give me this idea. Rao shared all my emails within his organization (United Nations) and told me how his colleagues were looking forward to seeing the book in print.

I would also like to thank Preeti Mehra for encouraging me to write and publish some of the stories in the book in *The Hindu Business Line*.

My sincere thanks to Chiki Sarkar, who took the time to read the book and give her valuable comments, Siddhesh Inamdar, whose inputs were extremely useful, and the many publishers who read my manuscript and did not discourage my efforts.

Preface

I quit my corporate job as a project manager with IBM in 2003, after working for fifteen years in the Information Technology (IT) industry. It was not a career change but a change in lifestyle that I was looking for. I was tired of the vicious rat race in the city, the pollution, the traffic and the chaos in everyday life. I was frustrated by the mechanical and insensitive city life and the blinkered or complete focus on earning more and more money.

I changed my life to become a farmer, which was unthinkable for me until a few years ago. It was not smooth sailing. I was a complete novice in this field, never having lived for even a single day in a village. It was a challenge to make the transition from the city to the village. Farming was a new skill that I had to learn from scratch, and unlike software or hardware there were no manuals or help buttons to guide me along. I had to learn the hard way by experimenting and trying out new things.

Preface

When I quit my job, a few friends at IBM wished to be in touch and stay updated about what was going on in my life. I started sending fortnightly email updates to them. What started as a small group soon grew to include relatives, acquaintances and more friends. I wrote to them in detail about how I was managing the transition and trying to break away from city life while learning different skills and a brand new profession. My narration always evoked sympathetic responses and stirred the emotions of my readers. Many of them were still in the corporate sector and deep within them, they had the desire to break away and do something more exciting.

In this era of globalization and a money economy, there are few who will give up the chase for big bucks. This is the time when hundreds are migrating from the villages to the cities looking for elusive jobs and secure incomes. I felt rather alienated by this rat race and isolated too in my attempt to give it all up and try and eke out a simple existence by farming. Some of my friends suggested that I should collate my experiences and make them into some sort of a book which people could read and enjoy. They were of the opinion that it would make for interesting reading and may prove to be an inspiration for many others who were on the threshold of such a 'reverse' migration.

At first I was extremely sceptical about this idea. I had never written more than a few words at a time, usually letters to my cousins as a child, or later, emails which were work related. A book was the last thing on my mind.

The driving force behind the book is my wife, Meena, who encouraged me and gave me the confidence to start writing down my experiences and feelings as I went about this transformation. During the monsoon in 2005, when I had some spare time as the work at the farm was not much, I decided to start writing.

I am thankful to Meena for having guided me in my first attempt to write a book. I would also like to acknowledge the tremendous encouragement from all my IBM colleagues and friends on the email list who inspired me to do this.

1

Where Am I Headed?

To Leave or Not to Leave

It was nearing six in the evening. I stared out of the blue tinted glass windows of my plush office at Bandra Kurla Complex, Mumbai, and watched the sun dip into the western sky. It was time to start the long journey back home through the rush hour traffic. Just as I was about to press the 'Turn Off' button on my IBM ThinkPad, Sriram, my senior colleague, tapped me on the shoulder and asked if I was in a hurry to leave.

Sriram is a soft-spoken, easy-going man who had been my boss for many years before he took on a different role in the organization. Years of working with him had taught me that he rarely asked a question without a reason. I quickly replied that I was not in a hurry and asked if there was anything I could do for him. He said, 'There is a conference call I would like you to attend at 20.00 hours. We can leave after that.'

This meant only one thing in IBM. It had to be a call from the Americans. Our head office was in the USA and due to the time difference, these conference calls were scheduled late night or in the early hours of the morning. We jokingly referred to them as 'con calls'. Of course, if Sriram was asking me to attend, it had to be something important.

I called my wife, Meena, at home and told her that I would be late again. Meena, then a freelance journalist, worked from home. She also did various projects and research work which took her into the remote villages of India. She was at that time working on a book on organic cotton. She had just returned from one of her trips in the morning and we had hardly met before I left for work. We had planned a quiet dinner at home but now that would have to wait.

I had two hours to kill before the call, so I grabbed a cup of coffee and went to the open space on the first floor to gape at the sunset and get some fresh air. If the con call meant a project offer, I would have to take a decision soon.

My cell phone rang suddenly, stirring me from my thoughts. It was Sriram trying to trace me. I rushed back to the conference room where he gave me a brief on the call we were to attend.

IBM was undertaking a conversion project with a third-party vendor in India and there was a huge budget earmarked for the exercise. The project was to be controlled out of the head office in USA and a lot of global attention was focused on its success. After screening many potential candidates within the organization, I had been selected to head the project for IBM, based in Mumbai. It was a tough project and they needed the best hands to handle it. The call that day was to introduce me to the team so that I would be part of the process right from the start of the project.

After a two-hour call where I was introduced to the team and also the entire project and its finer details, we left for our homes. I had reams of documents to read and understand before I got the hang of the entire project and its nuances.

As I squeezed my car into the never-ending stream of homeward-bound traffic, my thoughts were confused. I was working for one of the best companies in India with a superb project on hand and yet a part of me was not sure if that's what I really wanted. My mind was in turmoil. Hardly a week ago I had reconciled myself to the idea that I had to quit this organization. There were things I wanted to do with my life.

Changing Over

It was in 1996 that I had quit German Remedies and joined Tata Information Systems Limited (TISL), which was a joint venture between the TATA group of companies and IBM, the global computer giant.

My job at TISL had been a new experience for me. After years of working for the industry in the in-house computer division, I was for the first time on the other side of the table. I was a consultant. We used to implement projects for different companies and the work was very challenging. It was a high-pressure job and everything was time-bound.

Our office in Nariman Point was a small one and we had to share one desk with our boss, Sriram. He would always guide us and help us out whenever the need arose. We were working for a new organization and the infrastructure was poor. We would all take turns to use his computer to check our mails and complete various administrative tasks. Even though we had many problems, the atmosphere at work was excellent and it was a joy to report to office every day.

I enjoyed my new role and did it well too. I managed to bag a few awards within the company and also got a good raise in my salary. Within a couple of years IBM was to buy out TATA and we were to become a fully owned IBM company. I was soon working for a multinational company (MNC).

Working for an MNC and that, too, one of the best in the world was a heart-warming experience. The company looked after all the needs of the employees. We moved from our poky office in Nariman Point to a large one at Bandra Kurla Complex. The office furnishings were all according to international standards and so were the facilities. The employees were expected to be given all comforts so they would do their jobs rather than run around and waste energy in administrative matters.

Besides the material comforts they provided, our salaries were raised to global standards. The company had liberal leave policies and we were expected to take a minimum of ten days' leave every year. This was good for us as we could travel every year to a new location. We did exactly that and within a few years we had visited Bhutan, Nepal, the northeast states of India, and even managed a trip to Switzerland and Paris.

The job itself was extremely satisfying. We got to undertake software projects with some of the big business houses in India and implementing them was rewarding, especially since we followed some of the best global practices. The company also sent us abroad for training sessions. I was soon promoted as project manager and also certified by an international agency as a project management professional (PMP). There were few people in the country who were certified at that time and it was a big credit to our company.

Every project in IBM was considered a profit centre. We had to ensure that we stayed within the budget and made some profit at the end of the project. This was extremely challenging and it made the job all the more interesting. We had to look for ways of keeping the customer satisfied, for the focus of the company was very clear. Customer satisfaction was the foremost goal of the company. Every project was a tightrope walk, trying to balance the budget and the workload.

I seemed to have everything I desired, except the time to enjoy all of it. At IBM, there were no rules or pressure directly to perform. It was up to each employee to do their best. We did not have office timings or fixed leave or scheduled holidays. There were no machines to clock your in and out time. No one objected if you walked in late or left early. We were all provided with mobile phones and laptops so we could work from wherever we were located.

The company worked on the principle of 'pay by performance'. The sales personnel had targets to meet and the service personnel had utilization to achieve. Utilization was the time you spent on projects which were making money for the company. You were considered unutilized if you sat in office and were not assigned any project. Every person who achieved the target was given huge monetary incentives besides trips abroad on company expense as a holiday. This kind of a setup has a flip side to it. It sometimes leads to self-imposed stress and pressure to perform well and stay on top.

The connectivity and infrastructure ensured that each employee was always on call if so desired. You could not get away from work at any point of time. In spite of various attempts by the company to bring in the balance between work and family life there was always this big gap in giving quality time to the family. We even had family outings and parties to ensure that employees and their families had fun. These outings would start in earnest but after a couple of drinks and a round of housie, the conversation would veer towards business. It was as if you were on a call with the company 24/7.

It was not that all employees of the company were conscientious and under stress to perform. We had our own group of employees who were called the 'cc' group or the carbon copy group. These were the people who did nothing but lip service and whenever a deal was coming to a closure, they would ensure that their names appeared on the 'cc' of each mail. They made sure that they got credit for the deal.

These were the guys who spoke the most during the 'con calls'. There was nothing that could be done with them.

After working with IBM for seven years I started getting restless. Being a project manager meant that I had to spend a lot of time on site with the team to ensure that things went smoothly with the project. The travel and long absences from home was also getting on my nerves. There were many of my colleagues who also felt the same and were making efforts to find a new job.

It was during one of those informal chats over tea with my colleagues that it dawned on me that maybe I did not need a new job. I thought that another job would be as good as this one. What would be different in a new job? Maybe the office location and the people who worked there. I would still be doing the same thing. It was clear that everyone needed an expert. If I was good at project management, all other companies would also offer me the same role. I needed to rethink what I wanted to do.

A new job would mean more money but then we already had enough. What would we do with more money? Probably buy a bigger house and a bigger car. A bigger house would be good but it was not a necessity. Even if we did buy a new one what could change? The view outside our window would probably still be of an air conditioning duct or worse still a clothesline. It would also mean we were back into the debt cycle with a huge Equated Monthly Instalment (EMI) to pay every month.

I also wondered about the quality of work I was doing. I was putting all my effort to make some software work for a company which in turn paid IBM lots of money. Some of the money was handed over to me in the form of salary and perks while the bulk went out of the country into the pockets of the multinational corporations. The same money then came back to our country as debt or charity for some natural disaster or aid for government programmes.

The office atmosphere was great and the work ethic just wonderful. But I realized that work was becoming all-encompassing. There was hardly time for anything other than work. Besides, if I was on a project, I was away from home for long durations. I was thirty-seven years old and wondered if this is what I wished to do for the rest of my life.

I could not help but think of Mukta, one of my college friends, who had such great plans for her life which ended so abruptly with her death even before she was twenty-one. Life was short and uncertain. I felt it was better to do what I wanted to do when there was time and energy. Who knows what will happen in a few years?

I thought a lot and recalled all that Meena had told me of her various trips to villages during her research. Why shouldn't we live a life where we could grow our own food and live close to nature? We could shift to a small village where we built our little house and lived off the land and its produce. The idea was forming in my mind and I discussed it with Meena.

I reasoned with her that we did not need a lot of money to live. Especially if we were growing our own food, we would not need much. We reasoned that if we could produce the bulk of our calorie requirement from the land itself there was hardly any other expense we would incur. We would need to spend on some essentials that we could not produce like salt and some spices but that was not much. I had no idea how much was needed to live in a village but knew that it would not be as high as what we spent in the city. We did not have children and did not have to worry about their schooling and relocation. We had the additional advantage that we had no major liabilities. Our parents were independent and were not staying with us.

I discussed this with many of my colleagues and friends. I was pleasantly surprised at their reactions. There was not a single person I spoke to who thought it was a crazy or silly idea. They were unanimous in their view that this was an excellent idea and they all wished they could do the same thing. However, each one had a very good reason why he or she could not do something like this. It was either their children or their parents or a car loan or a housing loan that was stopping them. Every person I spoke to seemed to feel that they needed to get out of the system. It reflected the levels of frustration and stress in the city.

I asked all of them that if the desire to do something different was so strong and the stress levels so high, then why was it that they did not do something about it? The replies were downright silly or materialistic. I realized that it was insecurity and fear that prevented people from looking at things differently.

Besides, money played a major role in most replies. People were frightened that there would be no money and they would be left high and dry. I recalled my early days at Rallis India (my first job in 1989) when I got Rs 750 per month of which I gave Rs 200 to my mother for household expenses. In those days, too, at the end of the month I would be short of money. Things had not changed much over the years. My salary was in six figures but yet I was in the same predicament at the end of the month. It was only that now we had plastic money to bail us out.

I realized that money was never enough. As your income grew so did your needs. You travelled by bus and when you had more money you bought yourself a car. When you got some more money, you thought of a newer or a bigger car. It dawned upon me that the road you travelled to reach office was the same and the time of travel remained unchanged. It was immaterial whether you were in a Maruti 800 or a Honda City. It took the same time to reach office. It was more of a status symbol to have the latest gadgets and equipment when in reality you had lesser and lesser time to enjoy them.

To save time and keep you 'connected', you had more expensive gadgets that brought you closer to work at all times. Where was this leading? The more I thought on these lines the more I was convinced that I had to break out of this cycle. It was not that I was not fond of money or had no fear of the future. I reckoned that if you had to do something that had an element of risk in it then it was better to do it when you were young. You had more energy and the chances of retracing your steps were better. Some people I spoke to felt this was a retirement plan and they wondered why I was thinking of it at such a young age. I reasoned that one could never do this kind of work on a farm and take such a huge risk after retirement. It just would not be possible physically and financially. Besides I was not looking for retirement, I was looking for a new lifestyle.

Rajesh, my colleague at IBM with whom I had discussed this hundreds of times, advised me to check my financials before jumping into a venture of this magnitude. I took his advice and started working on my financial spreadsheet. I calculated all my liabilities and listed down our assets. I called up my head office and figured out how much money I would get from IBM if I quit. In true project-management style, I listed down our risks and the mitigation factors for each one if we did decide to take on this change. We spoke to some friends who had made the transition to living in villages and tried to figure out what monthly income was considered optimal to live in the village. I set myself a series of tasks that I would have to undertake before actually plunging into the operation.

Once I had done my homework, I shared it with Meena and some select colleagues in IBM and asked them to look at it critically. They all made some valid suggestions and changes to the structure and by the beginning of 2003, I had done a clear financial analysis with risks and had a mitigation plan ready. All I had to do was quit IBM and I would be ready with the money to start my new venture.

The plan was simple. With the money we got from IBM we would buy some land and build a small house to live in.

We would then till the land and live off it. We planned to grow all the items that we needed to survive. Maybe keep some poultry and cattle for milk, eggs and meat. On paper it all seemed great and possible to accomplish. At this stage I was raring to have a go at the idea. It was always the case with me that when I started on a new assignment or project, I jumped into it with total enthusiasm and energy. Meena was a bit cautious about the whole thing. She felt that I should quit IBM only after we had at least identified the land we intended to buy. That way we would be sure that our plan was in place and rolling before I quit.

The first step in our plan was to clear our outstanding liabilities. I closed my car loan and housing loan. I told my mother about my decision and also that I would not be able to send her any money, at least for some time. At first she seemed a little perturbed by my decision. After all, who in his right frame of mind would leave a cushy six-figure salary and venture into a life full of uncertainty and risks. In spite of her apprehensions, she was most understanding and assured me that she would manage without the contribution from me.

We made a list of expenses that we thought we would incur on a monthly basis at the farm. After considerable research and discussion, we figured that around Rs 6000 per month would be sufficient to stay in a village in reasonable comfort. We earmarked a portion of our savings to be put into a monthly income scheme. There was a small sum we set aside for our emergencies and the balance was what we had as capital to start our farm. We had no intentions of borrowing to set up the farm. Frankly, we were not sure if we could manage the repayment if we did that.

All the groundwork and the planning had been done. The whole project looked good on paper and I was convinced that I could swing it. The con call was a jolt. It posed a fresh dilemma as I was selected to head a prestigious project. To take it or not, that was the question. Suddenly my confidence waned. I was not sure where I was headed.

2

Search for an Alternative

It was around the early 2000s that the Indian economy started booming. The newly established outsourcing industry mushroomed and an entire swamp near our house was filled in and converted into a huge business place with towers full of Business Process Outsourcing (BPO) companies. Anyone with a slight English speaking skill could land a job in the 'call centres', as they were called, and would get a five-figure salary to start with. The automobile industry was opened up and every well-known brand from around the world was available. The financial sector was flush with surplus money and was doling out loans at astonishingly low rates to anyone who desired to buy a vehicle. Suddenly the roads seemed to have grown smaller and we had traffic snarls right in front of our home.

The suburbs were suddenly growing out of proportion and the city was becoming unbearable to live in. Our quiet leafy suburb at Goregaon in Mumbai was suddenly home to a large number of malls and multiplexes. Every morning, it was a torture to drive to work with the traffic snarls and one had to leave early to reach office on time. The pollution, the noise and the crowd was getting to us. We had to get out of the city like many others to breathe easy.

To try and get away from the chaos, often on weekends, we would visit Karjat (100 kilometres from Mumbai), where one of Meena's friends, Ghulu, had bought some land. It was located at the base of a mountain, an hour's drive from Karjat station. There was nothing around except a forest and a small village nearby. It was the most unbelievable place we had seen. Ghulu intended to build a small house on her plot of land and live there. We saw the place and decided that this was what we wanted: our own space outside the city, where we could go whenever we liked and breathe some fresh air in peace and solitude.

The place was close to the Bhimashankar range and it brought back pleasant memories of our trek there long ago when we were in college. We met up with the promoters and identified a plot we wished to buy. As a precautionary measure, Meena called up the forest department to check the authenticity of the documents. We were told that some of the land could actually belong to the forest department, but the settlement of these places had not yet been done. This made it a risky venture and no one could predict when the government would crack the whip and reclaim the land.

Though the deal did not work out, the seed had been planted in our heads. We now decided to go all out and find a place similar to the one we had seen in Karjat. Almost every weekend we would take the car and drive to the outskirts of the city looking for places to buy. As we went about our search we learnt quite a few things that we had to be careful about while buying land. There were numerous rules and regulations that had to be followed before one could own land in Maharashtra.

We searched for almost two years and were nowhere near getting the land we liked. Every time we would have a reason for not liking the land. Either the road was too far away or there was no view or there was no water. Our search for land had a keen following within IBM and every Monday, my colleagues would gather around and ask if we had found anything. They were all keen on buying some land and liked our idea of a weekend getaway.

My colleague Shrikant once commented, 'You will never buy land. It is the thrill of searching that keeps you going.' Maybe he was right; there was a certain thrill and excitement in just going out of the city every weekend looking for land. By the end of 2000, we had covered almost the whole of Karjat and the neighbouring areas. There was not a single broker or agent who did not know us.

It was not until a year later, the end of 2001, that we found a piece of land that seemed to satisfy all the criteria that we had in mind. It was a small hillock close to the road and there were indications of water close by. It was just on the outskirts of a small village and had lots of trees on it. The price was right and the papers seemed to be in order. We paid a small advance and started work on the survey. We thought we had found our dream land but a few days later we realized that we had been cheated. The owner called back and increased the price twofold. We could not afford the price and the land was not worth so much.

It was after a couple of trips that we got wind of the modus operandi. The area was well known for these kinds of deals. Once the advance was paid, the owners usually backtracked and never returned the money. They were scamsters who had no intention of selling the land and survived on looting unsuspecting city dwellers like us. We decided to move out of the area and put our search for land on hold in Karjat area. We had an unpleasant time getting the money back but we did manage after almost six months and several trips to the village.

First Interest in Farming

In 2002, Meena took up a research project as part of her freelance work to study organic cotton farming and make a compilation. She was to write a report on the subject which would later be published as a book. During the course of her research, she travelled to four states in the country and visited numerous farmers and agriculturists who were into organic farming with a focus on cotton. She often stayed with the farmers while she went about her research. Every time she returned from a trip to the rural areas, she would give graphic descriptions of their lives there. I would grill Meena after every trip to get as much detail as possible. It seemed like they lived a hard life but somehow they sounded happy and to me they seemed better off.

She would give clear descriptions of their houses, the clean air they breathed and their hard work, often at the mercy of nature. The food they cooked from the produce of their land and farm was simple but delicious. In this age where everyone was inundated with information on mono saturated fats, cholesterol, lipids, High Density Lipoprotein (HDL) and Low Density Lipoprotein (LDL), the villagers had their daily food with dollops of home-made ghee. Every time she returned she would curse the food we ate in the city, citing examples of the wonderful meals she had eaten in some far-off village.

At home there were intense debates with Meena on the rural and urban divide. I remember her coming back all excited from one of her trips to Gujarat. She had stayed there for nearly a month with various farmers. I made her repeat in minute detail everything she had experienced in these villages. I think she too was a bit surprised at my hunger for detail. I started thinking about this a lot. Why shouldn't we live a life where we could grow our own food and live closer to nature? We could shift to a village, build a small house and live off the land. The idea was forming in my mind and to be sure I kept discussing it often with Meena.

I remember one evening when, after a long conversation, I kept replaying her stories in my mind. I sat crouched on the low seating in my living room as the light of the cane lamp formed delicate patterns on the walls. It was a mellow moment, yet a decisive one and it was to change my life. Yes, that was it. For long we had toyed with the idea of a weekend getaway. All those searches in Karjat and even in Lonavla—that frantic hunt for a bungalow plot—were over now. The idea of living in a village, with a small house and a bit of land around it, appealed to me very much. Let me dispel the idea that it was an easy decision. But it came to me all of a sudden.

I was amazed at the stories of the lifestyle of these farmers though I had not actually seen any of them closely. The lives of the farmers, and not all of them were big landowners, seemed tough but it did have, for me, a certain charm and purpose compared to our fast, stressful city life. They worked hard but at the end of it, they were proud owners of produce that they could eat and relish for the rest of the year. Their lives seemed to move at a steady but slow pace and it was obvious, at a superficial level, that their stress levels were low, though low prices, agrarian distress and suicide were an ugly reality and formed a dismal backdrop to their lives. The pictures she had shot as part of her research were vivid in my mind and I dreamt of owning and living in a house that was similar to the ones in the village. Suddenly, our tiny 530 square feet flat in Mumbai became claustrophobic.

I also liked the concept of organic farming and the whole idea of growing food without using harmful chemicals and other inputs. I seemed to be convinced that villagers were healthier than us due to the clean and natural food they consumed. I read Meena's research and tried to understand the ill effects of chemicals and the need to stop using them. I studied about the presence of chemicals and toxins in our daily food and how they impacted our lives and health.

While Meena was totally in agreement and felt that it would be a wonderful change from the city life, she felt my decision was too sudden and based on hearsay. She too was not exactly thrilled with the city and hated the local travel, the pollution and the crowds. She agreed that it would be a welcome change from the city but she was a bit apprehensive about the financial part.

I had my own share of health problems to keep me worried. I always used to get nagging headaches, especially when we trekked. It usually vanished after a good rest and we attributed it to the sun. During a medical check-up, the doctors identified that I had slightly high blood pressure. There could be a heredity factor since both my parents had the same problem. During these tests, I was also diagnosed with diabetes and high cholesterol.

The doctors assured me that there was nothing to worry about and I needed to have a few medicines regularly to keep these under control. So here I was, barely into my thirties and already popping pills of various colours. I switched over to coffee and tea without sugar. Desserts were passed over and I adopted a strict lifestyle. Daily morning walks were a norm and visits to restaurants and bars were down to a bare minimum. Although things were under control, we did not have much faith in allopathic drugs and their long-term effects. I had seen how my parents had started on pills and finally were on insulin injections in a few years. We decided to look for alternative cures.

A good friend told us about Tibetan medicine and I decided to give it a shot. There was a Tibetan doctor who visited Mumbai every two months and I decided to pay her a visit. Within a couple of months, she had managed to control my blood pressure and bring it within reasonable limits. Despite my strict regimen, it was only the diabetes that she just could not bring down. She seemed perplexed that her medicines, which were renowned for reducing blood sugar, did not seem to have any effect on me. She kept telling me that this had to be because of stress.

The more I read the more I was convinced that the main cause of my illness was the food I consumed. I was shocked at the amount of chemicals that was used in growing vegetables and cereals, which constituted a large part of our diet. It was also evident that stress and the fast lifestyle that we followed were contributing to our general ill health. It was no wonder that the Tibetan medicines did not have any effect on me.

I was sure that if all of us changed our lifestyles and ate good food all our ailments and modern sicknesses would vanish from the earth. Many of my colleagues at IBM were already afflicted with diabetes, high blood pressure and chronic back problems. I realized that the stress of corporate life was taking a toll on everyone and was the cause of so much illness around us. It seemed like the best option left was to return to our roots and do what our ancestors did long ago. Grow our own food and live a stress-free, happy life in the village.

This may have seemed romantic and foolhardy at a time when the country was going through its worst agrarian crisis. Many justify the Green Revolution and its intensive use of chemicals and say that our country was saved from starvation and famine. However, I still read reports of malnutrition and starvation, which somehow has not vanished after the great Green Revolution. In addition there is another major problem—the terrifying impact of chemicals on nature and human beings.

Since the last decade, hundreds of farmers have been killing themselves, unable to repay their mounting debts. The government's pricing policy, lack of proper rural credit and poor minimum support prices were creating havoc. It was a vicious debt cycle ending in tragedy for the farmers. There was no end to it. For me there were lessons to be learnt from this. I realized that being debt-free and using zero inputs could help in cutting costs of production. Many farmers who had turned 'organic' were realizing the benefits of this policy. This would mean a major shift to self-reliance which somehow had slipped from the farmers' grasp, thanks to hybrid seeds, numerous subsidies and sops in the forms of pesticides and chemicals, while keeping prices of produce low. When I later visited organic farmers in Madhya Pradesh and Tamil Nadu this was confirmed. I was inspired by their pioneering efforts and felt vindicated that eventually I had decided not to use chemicals.

Search for an Alternative

Working out the Finance

Our plan was taking shape now. We would buy land in a village and try to live there. We would have to build a small house and have some basic infrastructure. The whole idea was not to generate money but to try and live off the land. We would have to grow all the essentials that we needed to eat and the balance, if any, could be sold. This meant that the farm would have to be a mix of horticulture and agriculture. We would have to grow fruit trees and other trees but at the same time leave enough room to grow vegetables, cereals, oilseeds, rice or wheat and other essentials. In case there was not enough cash generated to buy stuff, we would have to dig into our resources and savings.

The detailed spreadsheet I had made projected that with the money we would have left after setting up the farm, we would survive for another three years. Of course this was based on an assumption that Rs 6000 a month would be enough for us to live in a village. If we sold off all our assets, including the car and the house in Mumbai, then we would have enough money to survive for ten years before we were broke. This was assuming that there was no income from the land and we were dependant on the money we had saved all these years. It meant that we had three years to set up the place and turn it around before we started digging into our assets.

It was a tight budget.

The time was short.

We had no knowledge of farming.

We had never lived in a village before.

We had not yet found any land we liked.

Ah! This sounded like so many of the projects I had handled for IBM . . .

Everything was in place on paper for the transition. All we needed was a piece of land to make this come true. We started looking in earnest. It was just like earlier when we were searching for a weekend getaway except that our criteria had changed. Now, we were not looking at the scenery but at the soil. Was this good soil? Was there local labour available? Where was the water source for irrigation? What kind of trees grew around the place? What were the traditional crops in the area? And so on. Also, from one or two acres of land we were now looking at a larger piece of land, around 4–5 acres.

We were exploring areas around Dahanu, about 120 kilometres from Mumbai. There was a lot of land for sale in these areas but they were larger tracts of 15 plus acres which we could neither afford nor handle. We had to find something that was small and would fit into our budget. As days turned into weeks and then into months, we were nowhere close to buying land.

A Cinematic Encounter and the Final Plunge

It was around June 2003 that Meena accidentally met one of her old friends, Raajen Singh, in a library in Mumbai. Raajen had a small farm near Boisar and in the course of their conversation she mentioned that she was looking for a farm.

Raajen immediately put us in touch with a broker named Moru Valvi in the Nanivali village close to his farm. We spoke to Moru and he promised to show us land which was for sale around his village.

As scheduled, we landed up at his place on a Sunday morning so he could show us around. One of the first places we visited was a beautiful farm in the village of Peth. The land was on the banks of the River Surya with small hillocks in the distance. The total area was 4.5 acres which was just what we were looking for. It had a fence running along the perimeter and was fully overgrown. There were a few *chikoo* (sapodilla) trees and a small house at the centre of the field. It looked shabby and unkempt but seemed to have great potential. We liked the place at once but the owner was asking for around Rs 16 lakh, which was way beyond our budget.

Moru Valvi, or Moru Dada as everyone called him, our broker who was also the sarpanch of Nanivali, showed us a lot of land but nothing caught our fancy. We exchanged phone numbers and were assured by him that next week he would have more land to show to us.

We had not yet identified the land we wanted and nothing seemed to be moving positively. Also, I was finding it difficult to make time to go to the village looking for land regularly. Only Sundays were free and sometimes I had work to complete and we could not go. I realized that for things to happen I would have to concentrate more. Finally, a week before the American con call, I convinced Meena that the only way to take our project and dream forward was to quit IBM and concentrate on this full time. Though Meena understood the gravity of the situation, she still stuck to her stand that we needed to identify the land at least before I quit.

I realized it would be impossible to juggle the new project and look for land. The project was scheduled to start in December and my conscience did not permit me to start the project and then quit it midway. Finally, in the first week of November 2003, I decided to submit my resignation. It was a huge step for me. I was leaving the corporate world where I was supposedly doing well, and stepping into the unknown world of farming. I had no idea what it was to farm and live in a village. There was only one thing I was sure about. I felt encouraged by the fact that 65 per cent of our country lived in villages. There must be something going there, despite all the deaths, the gloom and the doomsday predictions.

At IBM, when one resigns, you have to fill in an electronic form and submit it to the system. On 4 November 2003, after reaching office, I called Meena at home to tell her that I was about to submit my resignation. Meena was bursting with news. A few minutes ago, she had got a call offering her a job with a newspaper in Mumbai. It was an amazing coincidence. This was something we had not expected. I pressed the submit button, taking this new development as a green signal.

3

Searching for Land

The Transition

It was a month of intense speculation at IBM after I had submitted my resignation. I had spent seven long years there and knew quite a lot of people within the organization. The phones were buzzing all through the day with calls from different IBM locations as many people wanted to know where I was going. I had told only a few friends about my venture. After discussing my current assignments and work, IBM decided to relieve me on 31 December 2003. It suited me fine. I would start the New Year on a clean slate.

I woke up early on 1 January 2004 to get ready for work. It suddenly dawned on me that I had nowhere to go. I was unemployed. I crept back into bed and tried to sleep longer. It was an inexplicable feeling. I was sleeping, knowing that I could do this for as long as I wanted. After working for fifteen years, now I did not have to rush and get ready to beat the traffic snarls while I drove to office. I did not have to match my shirt with the tie I wore that day nor did I have to gobble up my breakfast as I rushed down the stairs. I did not have to worry about conference calls or wonder if the proposal I had submitted was opened or not. The only scary thought was that at the end of the month, there would be no fat salary cheque.

It took a whole week for the impact of my action to sink in. It was one long week and sometimes there were moments of panic when I thought of what I had actually done. Before long I got over the initial tense moments and was back searching for land in earnest. It was no longer only on Sundays or holidays that I could go to the villages. I would go every two or three days and meet as many people as I could. They would take me across slushy fields and over tiny hillocks identifying pieces of land they claimed were for sale. I would shortlist a few and then Meena and I would see them together. If we liked any, we would move to the next stage of financial discussions. It was a slow, tedious process and the urgency was only for us. For the villagers, things moved at their own pace. This was the first lesson I was to learn in my new life as a farmer.

Time was not important. Everything had to be done at a steady pace in the villages. I would plan to meet someone at 2 p.m. in the afternoon only to find them walking in at 4 p.m. The explanation was simple; they had missed the bus. You miss one bus and the next one was only after two hours. At first I would pace up and down like a caged tiger till I realized that there was no hurry. I learnt to pace myself. I learnt to sit for two hours at a bus stop just watching the cattle go by and the dragonflies buzz around. I learnt that two days could mean anything from a week to a month. You rushed around with not a moment's rest only in the city.

I had still not found an area that I liked. There was something about buying land. When you reached the place and stood at the centre of the field you had to get a good feeling, that's what a friend told us once. We were still not getting that 'good feeling'. The land should talk to you, my friend had said. Every time we looked at a piece of land we thought was great, we would feel something was telling us not to buy it. We just had to trust our instincts.

I stopped taking the car all the way to the villages. It was too expensive, besides the price always went up the moment people saw a car. I travelled by train to Boisar and then took a bus. As I sat in the train on my way back after another fruitless search, I could not help but think of my past, the city and the transition I was trying to make. I had spent thirty-seven years in this great city of Bombay, now renamed Mumbai. The city remains the same no matter what you call it. In thirty-seven years they say the city grows on you. You become part of a well-oiled machinery where every nut, every screw, every cog goes about its task unmindful of everything else.

It must be the only place in the world where the 7.53 a.m. fast local train has more meaning than a bleeding person lying by the tracks, knocked down by a train as he was crossing the tracks, probably while running to catch the 7.49 a.m. double fast.

My first job in this city with Rallis India had its share of train travel to Churchgate. The daily discussions in the train would cover wide-ranging issues. How do we get the prized window seat? Was Holi on a Sunday (which meant one holiday less)? Will there be flooding of tracks in the monsoon? Or on how the outstation trains delayed the local trains coming into town. One would think that these outstation trains did not carry any people and were only being run to cause havoc to the city's commuting masses. At no point would the discussion centre on why people just could not make it for an earlier train. It was sacrilege to even suggest such a thing.

The great financial capital of India, Mumbai, is kind to all. During that time, there were numerous ways of making money here and there was a lot of money around. The service industry was growing and even fresh graduates were earning five-figure salaries. Discussions were changing in office canteens from train problems to cars. People talked about the latest mobile phones or the different cars available, their merits and demerits and, of course, the all-encompassing car loans. Many a precious hour has been spent on the interest percentage and the best deals in town. Spreadsheets were exchanged with the latest and most complex analysis on the loans available.

Salaries were zooming for the middle class and money was so much in surplus that people were contemplating buying a second house or a second car or better still a bigger house and a bigger car. All one had to do was be a part of this fantastic money-spewing machine called Mumbai.

A Different Journey

As I sat in the train watching the landscape blur past me, my thoughts went back to the day's events. At 7.50 a.m., the Borivali-bound local was almost empty at Goregaon station. As the train entered Kandivali station, there was a thundering noise of people jumping inside to catch seats. These were the 'return' guys, the ones who took the train to Borivali and returned in the same train to go to their destinations towards town. A group came menacingly towards us and rudely asked an old man to get up. The man was completely puzzled as he too had a ticket. As he tried to protest, a kind passenger advised him to quietly get up since this group always occupied the same seats. Welcome to the train gangs. These young, energetic, noisy, well-dressed guys, toting the latest mobile phones, smoking in the train despite the various signs put up, suddenly took over the entire compartment. All those who were not part of their elite group were pushed out, shoved and generally made fun of.

I left them behind at Borivali from where I had to catch the Ferozepur Janata Express to go to Boisar. I couldn't help notice a large crowd standing in between the tracks, rather than on the platform. As the train rumbled to a stop I realized why they were standing on the other side. Their friends inside were part of a different train mafia which did not let you open the door of the train on the side of the platform. Only people standing in between the tracks got in and these were the regulars or members of some group.

After repeated requests to the person sitting inside, the door on the platform side was opened just as the train started to leave the station. I managed to squeeze in and sat among another gang of barbarians. During the journey, most of the chatter was about their workplace, bitching about their bosses or planning elaborate vacations. All of them had the latest mobile phones and this was the main topic of conversation.

Newspapers were circulated so in effect one got to read at least 2–3 papers during the journey. During this trip, a group of municipal doctors sitting across had a problem in opening the cover of their new Nokia mobile phone, so I volunteered to help them. Once the mobile was opened, they went back to their chatter with not even a thanks mumbled. You did not exist and even if you did, it made no difference to them. My attempt to start a conversation with one of them was politely ignored. This close-knit family did not allow strangers. I quietly went back to reading my newspaper for the rest of the journey.

An hour later, walking along the road near Boisar station, looking for the bus stop, I asked for directions from an old man. He smiled at me and asked me to accompany him. A few yards away we reached the bus stop. There was no sign or bus shelter; it was just a tea shop where the bus stopped. As the state transport bus ground to a halt, a few passengers shouted out loudly: 'Burhanpur! Burhanpur!' This was for the benefit of all those who could not read the fading sign on the bus. The door opened and the crowd allowed the old man to get in first, followed by women and children. The young men were the last to hop into the bus goaded by the conductor screaming, '*Chala! Chala!* (Let's go! Let's go!)'. It was amazing to see the bus fill up without a shove or a push. As I settled down in the dusty rickety seat, I recalled the barbarians getting into the local and rudely moving the old man out. I couldn't help but wonder which one of the groups was really educated.

The young boy next to me got up to offer a seat to an old man. I was suddenly overcome with shame. Shouldn't I have done this? As if in repentance I shifted to make space for the boy to sit. To start a conversation, I asked him his educational qualifications. He replied proudly, 'Fourth standard pass.' I politely asked him why he did not study further. He said without a moment's hesitation, 'The school is in the next village and I do not have money to go there every day.' I thought of my education and the money that was spent by my parents.

The old man sitting next to me smiled and asked, 'Where from?' His name was Balkrishna Ravindra Gharat. Within a span of fifteen minutes, he told me his entire life story. I knew the name of each of his sons and grandsons at the end of it. He wrung out every bit of information about my family and myself too. At this point, I could not but think of the mobile-toting gang of doctors in the train who did not even have the courtesy to thank me for helping them out. Ghetto behaviour is common in the city and each individual lived in his or her own cocoon.

Turning my attention to the people in the bus I found a lot of women who looked as if they were off to work. They were either teachers at the village schools or social workers doing some survey of sorts. Everyone knew everyone, including the conductor and the driver. The driver stopped in front of a school or *basti* for the women to get off. Each one wished the other a good day before getting off the bus with promises to meet in the evening. At one place, one of the women requested the driver to switch off the engine. It seemed like a strange request, till she stuck her head out of the window and yelled at the children outside to go to school. It was the first time I saw a human school bell.

At Nanivali village, my destination, after the day's search for land was over, I sat at the village tea shop for a cup of tea. The shop was run by Konduram's wife who was a fountain of information. She knew everything that happened around the village and was up to date on all news. She grilled me for the half hour I sat there and wrung out every bit of personal detail that she could from me. I knew that this information would soon be spread across the village by her. A few trips later my fears were confirmed. As I walked into Ambeda a few villages ahead of Nanivali, a strange man asked me, 'Ravi Seth, so you are still looking for land?'

The discussions at the tea shop ranged from the current crops to some recent murder in the next village. That was just before the general elections and they were all awaiting directions from the village sarpanch on whom to vote for. They were all in awe of the Electronic Voting Machines (EVM) which, they were sure, would deliver a big electric shock if they pressed the wrong button. I did my bit, rather futilely, to convince them that this was not true.

At the tea shop, I asked the lady waiting with me when the next bus for Boisar was due. She smiled and said it would come. When I asked for the exact time, she explained in a very simple manner, 'It will come once it passes us and goes to the village ahead.' Such simple explanations compared to our concept of timing where a 7.53 and a 7.59 can make a huge difference in our lives. The return trip was finally in a lorry carrying construction material for the highway since the bus did not come.

As the train trundled into Borivali, my destination, I couldn't but think of the differences between the city and the village, a mere 100 kilometres away. The transition had just begun for me. Will the city throw me out or will I be able to throw out the city? Only time could tell. I couldn't help but wonder that maybe it was not 'transition' but 'transformation' that we all needed.

4

Land at Last

The Search Ends

A few weeks later, we got a call from Ravi Dharmameher of Charoti village. He said he had heard that we were looking for land and he was interested in selling his farm. I told him I would check it out on my next trip. To my surprise, he told me that I had already seen the land. It was the one at Peth village, the one we had seen in October, which was unaffordable. I told him that we did like the place but it was beyond our budget and we could not afford it. To this, he said, 'I keep the price high so I can ward off tourists who waste time. But I heard you are very keen and have been coming there very often.'

I was pleasantly surprised at the conversation. The very next day I rushed to his house at Charoti to discuss the price. Ravi Seth, as he was locally known, was a successful transporter who belonged to the Koli or fisherfolk community. He had bought the farm at a cheap price from someone but had no interest whatsoever in farming. He was willing to sell if it made him a neat profit. I returned home that day for the first time feeling that things were in fact moving in the direction we wished. He had not thrown me out of his house on hearing my offer but had asked for time to think. That was a good sign.

It was a tense week that passed. I would reach for the phone every day to call him and then stop. I kept telling myself, 'Don't show too much interest in the land. Let him call.' He did call after ten days, by which time I had bitten off most of my nails. He wanted to negotiate. The next day at the farm, he took me around, making it a point to stress various things which would jack up the price. Each time he pointed to something and praised it, I would poke holes in it. We spent half the day haggling over each point till we reached a happy agreement.

Two months after quitting my job, I had finally found the land I liked. I started to work on the tedious procedure to get the land registered in my name. The first task was to find a good lawyer who would not cheat. A difficult task but we did find one in Palghar who seemed to know the ropes besides being recommended by Raajen. He put a notice in the local paper stating that we intended to buy the land and in case anyone had objections they had to get back within fifteen days.

The next step was to get a No Objection Certificate (NOC) from the deputy collector's office which gave us permission to buy the land. Then we were to pay the stamp duty and get our agreement printed on stamp paper. After taking an appointment with the registrar, we were to pay the balance amount to the owner and get our agreement registered. The registered agreement would then be submitted to the local land records officer or 'talati' who would issue the relevant documents in my name. It sounded simple and straightforward to me.

I expressed my surprise to the lawyer at the simple procedure. He smiled as if he had an ace up his sleeve. He told me that we should take it step by step and I should start with the certificate first. I understood the reason for his smile later. I was in for a shock when I reached the old cottage in Dahanu where the revenue department of the government is located. I was greeted effusively by the clerk who even offered me tea before inquiring what I wanted. I told him that I wanted to apply for permission and an NOC to buy land. He gave me a dirty look as if I had committed a crime by mentioning the permission. He said, 'Please contact some local lawyer who will speak to my senior and all will be done.' A bit confused about this procedure, I went back to the lawyer who told me that getting the NOC was not easy as it had to be done by underhand means.

I was shocked when I heard this. I had not expected this at all. Besides, I was not willing to do something above the law. My lawyer patiently explained that this was the usual practice and if I wished to do it my way then I would have to go on alone. The only help he offered was to draw up an application for the permission. He told me to attach proof that either my ancestors or I owned agricultural land somewhere in the country or that I was a graduate in agriculture and then submit the form to the department concerned. He also added that he had little hope that I would get the permission without his intervention but wished me luck.

I discussed the matter with Meena. She supported my decision and said that under no circumstance would we resort to any underhand deals. It sounded like a great principled stand till we discovered that neither she nor my family had any land with them now. Of course neither of us were agriculture graduates. Years ago both our ancestors had been farmers but over a period of time we had lost control of all our land. It was either sold or taken over by people. I called my mother and she too could not remember any land that belonged to us. All she could recollect was that my father's father had owned huge tracts of land till the 1970s but no one had any idea what had happened to it after that.

It looked like our little venture was not even going to take off. I even thought of an agricultural course since at that point that seemed the only option. I bought all the relevant books on the Maharashtra Land and Revenue Act and pored over them, trying to figure a way out of the situation. Buried among the inane texts of law and acts was one special clause which mentioned that the collector had the authority and right to give permission to a person to buy agricultural land even if he did not possess any land currently. Armed with this clause I went and visited the collector at his Thane office.

The collector was an extremely nice gentleman who agreed the law existed but admitted that they never used it. He said it was a question of setting a precedent. If one person gets the permission, word spreads around and soon enough there would be hundreds lining up. He explained that the law was in place to avoid agricultural land being gobbled up by land sharks. I reasoned that he was right and it may be true that I intended to use the land for agriculture but how would the collector be sure of this? I returned home realizing that our only hope was to get some land record to prove we came from a generation of farmers or go back to college. The other option was to pay up the bribe to the gang of crooks and forget the whole procedure.

In March 2004, I decided to try my luck and embarked on a journey to my ancestral village somewhere in Kerala. I had never been there and knew nothing other than the name of the village. I called up my father's brother who lived in Kerala and asked him to help me. He too had no records of any land but at least knew where our village was located.

A few kilometres from Palakkad, on the way to Thrissur, was a small lane leading through rice fields to our village Kootala. I reached the village and looked around. There were only four houses and a temple left. The entire village had migrated to the city like my father and his family had. I knocked at the first house, explained to them who I was and the reason for my visit and solicited their help. They were not forthcoming at all and denied ever having heard of my family. I was summarily dismissed from there.

The next house was more receptive and the old lady of the house remembered my grandfather and his family. She had also vaguely heard of my father and the fact that he worked in Mumbai in the Railways. I was glad I had finally managed to connect with someone in the village and explained my situation to her. It was heart-warming to hear that she did remember my grandfather owning huge tracts of agricultural land near the village but she had no idea what had happened to it. She called up her son Kannan who worked a few miles away and explained the situation to him.

Kannan had worked abroad for a few years before he returned to his village to start a small business. He could understand what I was trying to do by getting into farming and going back to a village. His only lament was that I was doing it in Maharashtra and not in my own village. He was very helpful and promised to take me to the village record office the next day. He also explained the reaction of the first house I had been to. That was the family which had encroached on our ancestral property many years ago and I realized the reason for their animosity. They must have thought I was back to stake claim on the land. I had no such intention and made it clear to Kannan that I just needed a certificate from the officials.

The next day, we met an old man named Thambi who was a broker in his heyday. He had extensive records of the land in the village and was kind enough to look them up for us. He identified the land that my grandfather had owned and also gave us a survey number which was listed in his books. Armed with this information we reached the local land records office only to find that the number given did not exist in their records. It seemed that recently the government had repeated the survey and renumbered each piece of land. I was losing hope when I remembered that even if the new numbering system was introduced, the records would usually have the old survey number. How else would the officials connect the old documents to the new ones?

Excited, I requested the officer to look for the survey number we had in the old registers. That seemed like a lot of work and our friend was not very forthcoming. I pleaded with him and explained that if I did not get the records, I would be on the streets begging. I have no idea if it was my pleading or the gentle prodding of Kannan, but he gave in and started looking for the number in the old registers. I could not control my excitement when in ten minutes he looked up and said, 'Your grandfather has land in the village.' The land may have been encroached upon but in the records they were still in my grandfather's name.

My joy was short-lived for the records were all in Malayalam and I could not read a word of it. I was sure that no officer in Maharashtra would know Malayalam to read it. I requested them to translate it into English and give me a certificate with the relevant details. This was now beyond the jurisdiction of the officer and he ushered us into the office of the area incharge. After another round of detailed explanations of who I was and why I needed the certificate, the lady in question agreed to give it if the tahsildar of the area agreed. She claimed that she did not have the powers to decide on this matter.

We rushed to the tahsildar who coincidently happened to be from our neighbouring village. He heard my story and then without passing the buck, agreed to give permission. He made it clear that there was no official format for the certificate as these types of certificates were not usually given to anyone. We were told to draw up a draft of the certificate so he could clear its contents. We immediately got a typist who typed out the letter. He stamped his approval on it, wished me all the luck for my venture and packed us off. We went back to the local office and got the certificate signed and stamped by the lady in charge. My trip to Kerala was not wasted. I had managed to get the crucial certificate.

I quickly faxed the certificate to Meena in Mumbai so she could show it to the revenue officials at Dahanu and get a confirmation that it was enough to convince them that we had land in India. I waited for a day in Kerala hoping to get a quick confirmation. Unfortunately it was not to be so easy. Meena's visit to the revenue office in Dahanu was fruitless as there was no official present to confirm the document.

On my return, I rushed to Dahanu to meet the elusive clerk, Mr Z, who looked extremely upset that I had returned with a document stating that our family had land in Kerala. He accepted the document and the application and told me to return after a week. I was jubilant at this breakthrough with the bureaucracy. I was sure that now I would get the permission and would soon be able to close the transaction and own the land.

My joy was short-lived. When I returned the next week to Mr Z, he informed me with a grin on his face that the papers were with the 'sahib'. I asked him who the 'sahib' was and he said the *pranth* or the deputy collector of the region. I was told that he was a busy man and not easy to meet but surely my papers would be passed next week. Of course, he informed me that if I was in a hurry I should just meet his senior, Mr X, who would sort out things for me. I knew meeting Mr X would put me in an uncomfortable situation.

I kept visiting the office every week only to be told that the paper had not yet returned from the 'sahib' and maybe I should try my luck the week later. I kept asking Mr Z if I could meet with this 'sahib' and plead my case since it was now almost two months since I had applied and even Ravi, the land owner, was calling up as he needed the rest of the money. Finally I was informed that the 'sahib' would meet me on any day except Wednesday in the morning. But I would have to come and try my luck as he was a very busy man.

I went a couple of mornings only to find no one at the office and the elusive 'sahib' missing. Each time I was informed that he was on tour or out on the field or had gone to Thane. I had a sinking feeling that they were lying and were just not letting me meet the deputy collector. I decided that I had to try out some other way to meet him.

From the next day onwards I went to Dahanu by the morning train and instead of going to the office, went and sat at the tea stall just outside the revenue office. I befriended the small boy who delivered tea to the office and told him to tell me if he ever saw the deputy collector in office. I would sit till evening at the tea stall and return by the evening train. This went on every day for a week except on Wednesday when I knew there was no chance of meeting the deputy collector officially. For one whole week there was no sign of the 'sahib'. It did seem that the deputy collector was a busy man and kept touring all the time. I even met some other poor souls like me who were doing the rounds for their work and were being asked to pay up.

The tea stall owner, Hari Om, was a kind man who asked me after a week why I was sitting at his tea stall every day. He said, 'I don't mind if you sit all day since you eat and drink tea here, but I am just wondering what you want.' I explained to him that I needed to meet the deputy collector and was trying my best to catch him without the department officer's knowledge. He calmly told me that the deputy collector had come to office on Wednesday. I explained to him that I had been advised by the officers not to come on Wednesday as he did not meet the public on that day. He almost rolled on the ground with laughter while telling me that I had been taken for a ride. He said, 'Sahib meets the general public only on Wednesday and not any other day.' I seethed with anger at Mr Z's masterly deception.

I told Meena about the whole episode and how the officers had taken me for a ride. She was livid. We decided that there was only one way out of this mess and that was to catch the deputy collector directly. Within a day or two she managed to get the contact number of the deputy collector from some friends in Dahanu and called him up. She explained to him that we needed to meet up with him for some land matter and requested an appointment. We were surprised when he gave the appointment for the following Wednesday.

Next week on Wednesday, I reached the revenue office early. I waited outside at Hari Om's tea stall. A while later Hari Om pointed to the official jeep of the deputy collector and said, 'There comes your sahib.' I called him up on his mobile after ten minutes and asked him when I could meet him. He paused for a moment before saying, 'Why don't you drop in now if you are in Dahanu?' I barged into the office and ran up the stairs to the first floor where his office was located. I gave my card to the sepoy and waited outside. After what seemed like forever but was actually five minutes, I was ushered into his office.

The pranth was a young man in his late thirties who smiled kindly at me and asked me to be seated. He completed the paperwork he was doing and then asked me what the matter was. I explained the entire story to him, leaving out portions where I had been asked to contact Mr X. He listened to me patiently, went through the documents I gave him and then called for Mr Z.

On seeing me seated opposite the 'sahib', Mr Z almost burst a blood vessel but managed to maintain his composure. The deputy collector thrust my papers at him and asked him why there was a delay in processing them. Mr Z mumbled and fumbled before informing his boss that the papers were in the drawer of some other officer who was on leave and hence the delay. The deputy collector told him to process the papers at the earliest and dismissed him. The pranth had an IBM desktop on his table and we started talking about the computer and some minor problems he had with it. I helped him out with the settings and he seemed pleased with the solution I offered him. We chatted for some time over a cup of tea and I left with assurances from him that my work would be done soon.

As I triumphantly walked down the stairs I saw a purple-faced Mr Z waiting for me at the foot of the stairs. He rudely asked me why I had gone to see his boss. I calmly told him that it was he himself who had said the papers were with the boss and so I went to ask about it. I asked him when I should return to collect my certificate. He said next week. I gently told him that he would have to be more specific. He muttered Tuesday and walked away. The next Tuesday I landed up at the office and a sullen Mr Z handed the certificate to me.

He was so shameless that after giving me the certificate, he asked me if I would part with some money as a tip. I replied in the negative and told him that he was getting his salary which should be enough for him to survive. To this he replied that their salaries had not come for the last three months and he was a poor man. This was now downright begging for money. I asked him if I should go and talk to the deputy collector and inform him that his staff were going hungry as their salaries were not being paid on time. He was quiet after that and I walked out of the corrupt revenue office with the NOC in my hand.

It was the last week of May 2004. It had taken me almost four months and innumerable trips to Dahanu to get the certificate. Yet, it was worth it. We had not paid a single paisa to get this work done. It just meant that with a lot of effort and patience it was possible to beat corruption, but that was a luxury not everyone had.

The lawyer could not believe that we did not pay a bribe. Finally, on 3 June 2004, we managed to get the agreement registered after paying the stamp duty and fee. Immediately after the registration, I rushed to the local talati's office and filed an application to change the land records to my name. After the long wait, I would finally be the proud owner of land.

At least that's what I thought, till I met the local official who had to sign the papers before the land was transferred to my name. He was a burly man with bloodshot eyes who calmly told me that he would sign the papers only if I complied with his demands. I told him I had no intention of doing so, and had managed to get all the permissions without resorting to any underhand means.

He was persistent and asked me to return after a week. Since he knew my wife was a journalist, he told me he was giving me a huge markdown for the final signature. I told him that I would return the next day when his head was clear and we could discuss this further.

The next day I reached his office early. As soon as he saw me, his eyes lit up and he started shouting, '*Ala re! Ala re!* (He has come! He has come!).' I tried my best for fifteen days to get the document signed by him. It was a wasted effort. Finally, I gave up and agreed. He honoured his part of the deal and I got the papers in ten minutes flat.

I felt extremely ashamed that I had succumbed to his demands to get the work done. My guilt grew even more when a year later I heard that he had been admitted to a hospital for a liver ailment. Somehow, I felt I was also responsible for what had happened to him.

It dawned on me that in the city we lived such cocooned lives that it was impossible to dream that just a few kilometres away the situation would be so different. Nothing moved here without bribes and there was no recourse available either. In the thirty years that I spent in the city, I had never encountered such red tape and bureaucracy that I had seen in the past four months. I realized that I would have to live with it and this would be a part of my life from now on.

My New Address

House number 752, Peth village, Dahanu taluk, Palghar district, Maharashtra—that was my new postal address. It is right at the end of Dahanu taluk, bordering Palghar taluk. There are eighty houses, mostly occupied by the Kunbis, with only a few Warli families. It is believed that the Kunbis were actually caretakers of the nearby Asherigad Fort built by the Maratha king Shivaji. In return for their services, he gifted them a lot of land in the surrounding area. That also probably accounted for the small number of Adivasis, the original inhabitants of the area, in our village. Since our village is too small, it does not have its own *gram* panchayat but is part of the group panchayat of Tawa.

The village lies between the perennial River Surya and one of its small tributaries. The two meet at the western end of the village and the sangam, as it's called, is particularly charming during the monsoon when the water gushes down in a series of small but spectacular waterfalls. On a clear day, you can see the pinnacle of the Mahalaxmi mountain beyond the river, which has one temple at its base and another one on top. The temple on top is inside a natural cavern and one has to crawl on all fours to reach the sanctum sanctorum.

On the southern side of the village you can see the fort of Asherigad at a distance. Several years ago, we had trekked to the fort which has a beautiful plateau on top, some ruins of the fort walls and cannons. I had never imagined that one day

Land at Last

I would be staying close to it and could look at it whenever I chose to. Tigers are believed to have roamed the area once, though now the wildlife is reduced to wild boars.

My farm is right at the northern tip of the village next to the River Surya. The road to the farm passes through the village past all the houses and lush paddy fields. As you cross the last house, that of Lahu Kaka, our nearest neighbour, the road takes a sharp turn to the left and passes through a thick private forest. About 200 metres from this turn you suddenly come to a clearing with an iron gate. That's the entrance. The approach road was a muddy track and during the monsoons it was not motorable. I kept putting some gravel each year and now the road is motorable even during the rainy season.

Midway between the gate and the house is an imposing jamun tree. In May it yields a slightly sour fruit which makes for excellent liquor brewed by the local women. From the entrance you can see the field neatly divided into four sections with the house at the centre. The four sections have chikoo trees planted by the earlier owner. Coconut, young teak, bamboo and some eucalyptus trees form a fence of sorts around the farm. Beyond these trees is a barbed wire fence to keep the cattle out. The fence was not exactly in good condition and there were places where the barbed wire had broken and access to the farm was possible. I had to repair and strengthen it later.

After buying the land, in the monsoon, I planted different fruit trees along the gravel path towards the house. All the trees started yielding fruit within five years, apart from the *amla* (gooseberry), which started eight years later. I planted four different varieties of mango, gooseberry, jamun (sweeter ones), love apples, lychee and coconut trees. As you come close to the house the first thing that you see are the bright pink bougainvillea creepers over the front porch. The house is a small structure with two huge mango trees on either side casting a soothing shadow. The structure is made of brick and cement with a sloping asbestos roof which is covered with small bundles of straw to keep the heat down. It was much later in 2015 that we replaced the asbestos sheets with a Reinforced Cement Concrete (RCC) roof.

The front porch has two platforms on either side on which one person can easily sit or lie down. There is a small swing at one end of the porch for lounging on. At night the shadows on the jamun tree give it an eerie human look, as if a huge guard is watching over us. On one side of the house is a vegetable patch with small mounds for root crops like carrot and beetroot, and radish during winter. Next to it are rows of vegetables of the season like brinjal and ladyfinger. The vegetable patch is lined with lemon trees. At the end of the vegetable plot is the fruit plot with lots of banana and papaya plants. Over the years I have collected around five varieties of banana plants. We have the one that is used for cooking, the elaichi ones (small yellow), a red variety from Coimbatore and of course the usual green ones that turned yellow when ripe. Next to it is the creeper section with a bamboo structure which allows the plants to climb up. Depending on the season this section would have runner beans, string beans, bitter gourd or ridge gourd.

The back of the house has a porch similar to the one in the front. It is a nice place to sit in the evenings and shoot the breeze. A small gate leads down to the banks of the river and sitting on the rocks by the river or swimming is another happy way to spend an evening. We later built a small *chapra* near the river to spend the evening.

The house itself was a small, modest two-room structure. We got it reinforced and built a small toilet inside. On entering the house the first room, or the living room in city parlance, has a simple divan, a writing table, a couple of chairs and a wooden cot. The bedroom is sparse with a cot in the middle and a couple of metal cupboards for our belongings. A decently equipped kitchen with a few vessels and a gas stove makes it comfortable. We added another bedroom to the existing house and renovated the kitchen into a modular one in 2015.

The nearest railway station is Boisar and there are buses every one or two hours to the village, which is considered regular. The bus stop is right on the outskirts of the village and it is a good fifteen-minute walk from the farm. Years ago the bus used to actually pass through the village till the private jeep and auto drivers managed to stop it. Even to go towards Dahanu one has to take a jeep or auto. Though officially they are permitted to carry only three passengers I have never seen any of them leave the place with less than six. The roads towards Boisar or Dahanu were in a condition that cannot be described. The potholes were big enough to accommodate any small car and it was like taking a boat ride. Many attempts were made to repair the roads but each time it lasted only for a few months. After every monsoon the potholes appeared again. It seemed that the officials took so much money as bribes that the contractors had no option but to cheat on the materials used for construction.

The Surya canal project brought irrigation to our village around twenty years ago and now people grow two crops every year. Being on the outskirts of the village, our land does not have access to canal water. The monsoon crop is of course paddy and during winter, paddy again or groundnut. All the villagers use urea and chemical sprays. The older generation remembers when rice was grown without any chemicals and pests were not such a problem. They have been growing crops twice a year for many years now and the land has had no time to recover from this continuous activity. Each year a section of the village screams for a break from the canal water but it happened only in 2010 when they did take a break.

It was a sheer coincidence that the land was exactly how we had imagined it would be. We did have a beautiful view on all four sides of the farm except for some ugly high-tension wires on the east side. If you wake up early you can hear a variety of bird calls and see a couple of crow pheasants who live on the outer edge of the farm. There are a large number of butterflies too and I have photographed some really exotic ones, including a bright, fern-like slug caterpillar. During the stay at the farm over the years we have spotted the Atlas moth and the luna moth. Winter mornings are a delight as the ground is swept by a faint mist and the leaves and grass glisten with dew. Fireflies can be seen before the monsoon and sometimes they even come into the house.

One of my favourite past-times, when I get the time, is to go down and sit by the river. The evenings especially are a treat with the setting sun covering the water with a gentle orange glow. The opposite bank is higher and it slopes down to the water. There are some hills behind. Often I would see cattle going back home, women washing clothes or schoolchildren frolicking in the water. It is an idyllic setting and makes me wonder if it's real. I always did love the outdoors as we call it. In the unlikely environment of a suburban railway colony, as a child, I had managed to grow a lot of plants and trees and often our games would involve a lot of tree climbing, something unthinkable in this age of computer games and 24/7 television.

5

Early Lesson in Farming

The First Crop

It was April 2004. I stood in the middle of the lush green field of moong (green gram) and looked around me. It was just before sunrise and the sky was turning a bright orange. The ground was damp and the leaves were shining with dew. My bare feet were muddy as I walked around gingerly, inspecting the plants.

Around me were rows of chikoo trees and below a dense foliage of moong. At that point, I could not have asked for anything more. The moong plants, not more than two feet tall, had green pods hanging out. The pods were not yet ripe and there was a light fuzz growing on them. There was still some time before the harvest. I felt exhilarated.

I stood watching the sun rise above the towering trees across the fence and slowly made my way back to the house, a white structure in the middle of this greenery. I could not

Early Lesson in Farming

believe that I was the owner of this land and that I was looking at my first crop as a farmer.

After I had paid the advance money for the land, I thought I would have some time to get familiar with farming. But Moru Dada, the broker who got us the land, had other ideas. He was keen that we plant moong at once. I was not prepared for this. I was still reading books and trying to figure out what we could sow and how we should go about it. Moru Dada was quite firm. He said the season was right for sowing moong and the best seeds were available in Surat in the adjacent state of Gujarat.

I was plunging headlong into something I was little prepared for. After all I had not even got the land transferred to my name and it had been only two months since I quit my corporate job. I realized that since I was keen on becoming a farmer, this was not a moment too soon. I made a quick trip to Surat and bought around 10 kilograms of moong. Moru Dada arranged to have our neighbour in the village, Baban Desai, help us on the land. He did not stay at the farm but came every day to help. Moru Dada rented his tractor to plough the land and quickly planted moong all over the place. The idea was that even if we did not get any harvest, we would still have some green cover which we could use to mulch the soil. I had started reading some books on organic farming and was picking up some stuff from the Internet too.

A few days later, we were overjoyed to see tiny green leaves. I had never seen moong growing before and was thrilled at the sight. It was the same thrill I had felt as a young boy when

I saw the first of the hibiscus I had planted bloom at the Railway Quarters in Vile Parle in Mumbai. I was grateful to have taken Moru's advice.

The next thing Moru Dada wanted to do was spray some pesticide on the plants. He claimed that it would give a higher yield. This was something we did not want to do. We were clear that we would not use any chemicals and tried to explain it to him. He reacted as if we had suggested hara-kiri. It took a lot of convincing to ensure that Moru Dada and his friends did not use any chemicals on the farm. They refused to understand how crops could grow without sprays.

We tried our best to explain to them that nature would do her job even without us interfering with poisonous chemicals. There were moments when we felt that maybe they were right, especially since we did not have any experience and were relying on what we had read in books or what Meena had researched for a book on organic cotton.

Contrary to what everyone had told us, nature did her job and she needed no bribes to get the work done. Soon it was harvest time and we managed a respectable 300 kilograms. An awful lot of moong and with it a lot of confidence. Now I was certain the land was fertile and that it was possible to grow crops without chemicals. It was a major morale booster.

I was terribly excited as this was the first crop from the land, even before actually paying for it. It had been only five months since I quit my corporate job at IBM and I had a decent first crop. I felt the transition was promising—from microchips to moong. We distributed a large portion of the moong to friends, relatives, neighbours and anyone who remotely expressed an interest. Even after the extensive distribution, we were left with almost 200 kilograms and there was no option but to sell it off in the market.

Selling Moong

We had an early lesson in farming after the first moong harvest. I went to the local grocer in Goregaon, Mumbai, and asked if he would like to buy the moong from us. He examined the moong in detail, took some samples for his home and then a couple of days later offered to pay us Rs 12 per kilo. We could not believe our ears for he was selling the same thing at Rs 30 per kilo. Why this massive difference? He gave us some vague explanation of how he had to keep the inventory for some time and it did not work out otherwise. Besides he got his moong from the distributor at the same rate. We were in for another harsh lesson in farming. The price we paid at the local grocer never indicated what the farmer got for growing the crop.

We tried to reason with him that this was completely organic and we did not use any pesticides or chemicals. He looked blankly at us and said, 'So what? Does this make your moong any different?' We gave up and decided that maybe we needed to go to the right market if we expected to get some appreciation for our organic produce. We contacted a vendor who specialized in organic food and offered our crop to them. They were very gracious and offered to pay Rs 17 per kilo since it was organic but we would have to deliver it to their godown.

With nothing else on hand and no idea what we could do with the load of moong we had, we decided to go ahead. We were also worried the moong would spoil, so we reluctantly agreed and went and delivered the lot. It was only after a month later and numerous calls to the person concerned that we managed to get paid for it. When the money arrived it was calculated at Rs 16 per kilo. I immediately called back and was informed that our moong had to be cleaned further and hence they had cut a rupee from the original agreed price. We quietly pocketed the money, thanking our stars that we had at least got more than the local grocer. We were in for a shock the next month when we got the rate list from the same organization, which listed our organic moong at Rs 38 per kilo.

So that was the bitter truth. All the sowing, watering, weeding, harvesting and cleaning are done at the farm. The cost of labour, seeds, water and electricity is borne by the farmer. But the bulk of the profits go to the trader who just packs the material and charges a bomb from unsuspecting city dwellers. We decided that this was the first and last of our produce that would go through the trader. The next time onwards we would do the selling ourselves even if it killed us.

Once the moong had been harvested, it was time for the monsoon season. By this time we had completed the paperwork and officially owned the land. The village sowed rice using the traditional method of growing tiny saplings on a mother bed and then transplanting them. We did not have so many people to work on the farm nor the resources.

Early Lesson in Farming

I had read a lot about farming, including the famous 'do nothing' method of agriculture by Masanobu Fukuoka. I was convinced that the current practice of sowing rice was labour-intensive and relied too much on external inputs. I was sure that if I wanted to make the farm sustainable and live within my means I would have to work out methods of agriculture that used very less or no external inputs.

It was an uphill task convincing the villagers and Baban that we wanted to do things differently and try out new ideas. Besides we kept repeating that we would not use any chemicals on the farm. This was received with much amusement and I would find strangers stopping me on the way and asking me if it was true.

I was just six months into farming and had no idea when to sow rice. Besides I was also having a problem getting labour and bulls from the village. As a result, I seemed to have missed the opportune time. I had to rush to the nearest agricultural institute in Kosbad and ask for advice. Unfortunately, the institute also promoted chemical farming and they clearly told me that they would not guarantee good yields using organic practices and it was my responsibility if it failed. They advised me to plant a particular variety of rice that would mature in ninety days, which was faster than the normal varieties. By using this variety, I would manage to overcome the delay.

I got the prescribed rice, GR4, from a shop at Dahanu and broadcast it across the field. We had got the land ploughed using a traditional plough rather than a tractor or tiller which was expensive. There was no question of spraying any chemical or putting urea in the field. I had a large number of visitors from the village who came to see the wonder rice that grows without any external input. Besides, they had never heard of anyone just broadcasting the seeds across the field.

We harvested about 100 kilograms, which was far below average. We realized that we would have to fine-tune the process next time. Obviously, we had done something wrong. We also realized that it was one thing to read and listen to people and another to actually practise what we read. Farming is something which does not come with a well-written manual and a convenient 'F1' or help button like most software. It has to be practised for years and learnt directly from nature itself. What worked for one farmer would prove detrimental to another. The parameters for each field and location varied.

This is also one of the major reasons why various research and experiments conducted by scientists at institutes sometimes fail when rolled out into the fields. Under a controlled environment, with a small area to be monitored, the results are easier to achieve while on acres of land open to the uncertainties of nature, the results can be extremely varied.

We learnt the hard way that it was not enough just to broadcast rice all over the field and expect a good harvest. We had to innovate and experiment to find the right method which suited the soil and environmental conditions. We would have to try year after year to reach some level of efficiency. We were convinced of one thing though, that there was no need to use chemicals and pesticides. We only had to learn how to harness the immense potential of available resources to our benefit. This was something that would not happen overnight and we had a long learning curve to pass through.

We decided to try and sell the little rice we had harvested directly. Our earlier experience had taught us a lesson and we decided to ignore the traders or the so-called organic shops. Meena and I sent text messages to all our friends. It was a pleasant surprise when within three days the rice was completely sold out. We suddenly realized that there were people who were interested in organic produce and willing to buy it from us. We could do away with the middlemen but we had the additional task of packing and delivering the material to the consumers.

We had charged Rs 20 per kilo and much later people called up and informed us that they were willing to pay even up to Rs 30. We had not got the rice polished but just dehusked and cleaned. It looked a bit brown in colour but the taste was exceptional, though the grains were not even. The proceeds from the sale of the rice did not even cover the cost of production. I did feel tense and worried that we were making a loss, considering that we had no other income other than Meena's salary from her job at *The Hindu*.

During the first few weeks after the harvest, I panicked and even contemplated rushing back to the city and looking for another corporate job. I had bouts of depression and went into long periods of mourning thinking of the foolish move I had made. I kept thinking that I had invested all my hard-earned money into the venture and was not even getting enough returns. I would probably have got more money just keeping the money in a bank or a fixed deposit. It was only after a number of meaningful discussions with Meena and reading a lot of books that I realized that things were not as bad as they seemed. It was just a question of time.

We reasoned that even though we had not made any profit from the rice crop, we had not dug into our capital. We were still living within our budget. It had hardly been a year since we started this venture. There is nothing common between software and farming and it would take me some time to learn the nuances of this new skill. It would be foolish to conclude that I had failed in this venture without giving it the right amount of time.

With renewed enthusiasm, I returned to my farming activities. I started planting vegetables around the farm. Everything was in small quantities though many people approached me with proposals to plant a single vegetable across the land. They were willing to share the profits with me equally. We were clear that we would not opt for monocropping as it would be both a risk and a bad thing for the soil. I was even approached by an exporter from Mumbai who proposed that he would buy all the produce from the farm and pay me 30 per cent premium over the prevalent market rate. He was exporting the fruits and vegetables to New Zealand and Canada where there was a high demand for organically grown crops. It was a lucrative proposal but we could not get around to agreeing to it. I felt it was better if we could sell our produce locally.

Early Lesson in Farming

Anyway, the quantity from the farm was not enough for bulk sales. I was happier to barter the produce with the villagers around who gave me vegetables I did not have on the farm in return. It was a good arrangement as we got to eat a variety of vegetables. I thought it would only be a matter of time before we almost stopped buying from the market and were completely eating what we grew on the farm. But that has not been the case—over the years we found that not all vegetables grew well at the farm and we could grow only local, seasonal vegetables.

We also observed that after eating the vegetables grown at the farm, our palates sometimes did not accept the vegetables available in the city market. Most of the vegetables we could buy in Mumbai were at least three days old. It is a different experience to go to the vegetable patch and pluck just what we wanted.

I had read up a great deal on organic farming and its benefits. I had a lot of book knowledge but still yearned to see some real fields that worked on the principle of organic farming. I made a trip to visit some farmers in the nearby state of Madhya Pradesh. I also visited some famous farms around our area which practised farming without chemicals. I made a fruitful trip to Tamil Nadu to attend a training session with some veteran organic farmers. It was a joy to see so many like-minded people who had switched over from chemical farming to organic farming. The trip made me feel confident that we had taken the right step in opting for organic farming and it would not be long before we would reap the benefits from our land.

I also visited the Late Bhaskar Save, who had his natural farm in nearby Umbargaon. It was a treat to sit and listen to him talk about natural farming and the ill effects of chemicals. He had switched over to natural farming way back in 1960 and the legendary Masanobu Fukuoka had visited his farm in 1988. During one of our informal chats he said something that would remain with me forever. He said, 'Nature will only provide for your needs, not for your greed. Farming is not a manufacturing business; it is a way of life.'

I sheepishly told him how I too had fallen for the greed factor. I had planted a lemon tree in the first year itself. The tree had grown well but there were no fruits. In my desire to get fruits I had called up all kinds of people, from scientists to village elders to nursery owners, trying to find ways of making the tree fruit. They gave various suggestions which I diligently tried on the tree for four years.

This was the advice they gave me:

- Stop watering after the monsoon: This was done for two seasons and nothing came of it.
- Cut a small one-inch portion from the main stem: This resulted in the entire tree turning yellow and almost dying. The amount of flak I received from Meena for this experiment cannot be documented.
- Dig pits around the tree and fill with ash and compost: This resulted in the tree growing faster and has reached almost twelve feet.

- Bury the fresh entrails of a goat: Got the entrails from the village and buried it without any change in the tree's behaviour.
- Tie a chappal to the tree: This was done but the chappals were taken away by some thieves.

Bhaskar smiled at me when he heard what I had done to the tree and said, 'Go back and apologize to the tree. She will fruit when she is ready.' I returned to the farm and asked for forgiveness from the tree for torturing it. Finally, after seven years of planting it, the tree started giving lemons. The lemons were the size of cricket balls. It was nature's way of accepting my apology.

During my various trips I made copious notes on whatever information I gathered and maintained a diary. This was extremely useful when some kind of pest attacked our crop and we had to find a non-toxic solution to it. I observed that by adding the leaves of the common calotropis plant to the base of the coconut trees, the yield was better. The nuts grew bigger and were sweeter than before. When we had bought the land we had got only three coconuts from our trees and a year later the same trees had more than two dozen each.

I also had back copies of the *Honeybee* magazine with me, which was a treasure chest of solutions to numerous problems. *Honeybee* is a magazine published by the Society for Research and Initiatives for Sustainable Technologies and Institutions (SRISTI), a non-governmental organization which was set up in 1993 to strengthen the capacity of grassroots inventors and innovators.

By now, I was starting to enjoy my farming experience. We were yet to generate any income from the farm. All we got was a lot of food to consume. These were grown organically and were extremely tasty. I was reminded of the fruits I ate as a child which Patel uncle, my neighbour, gave me from his garden. Our fruits too tasted different from the ones we bought from the market.

It was now clear that farming did not generate much income. If we did aspire for a lot of cash we would have to grow large quantities of crops. It was viable to transport the produce to Mumbai and sell it only if the output was huge. The small quantities that we produced were sufficient only for our consumption and maybe for some barter at the village or small local sales. I had reconciled to the fact that money was not important if you wished to live in the village and tried to live off the land. I realized that it was more important that we managed to grow all the crops that we needed at the farm rather than concentrate on one crop that gave us cash income.

A few months later, we had harvested groundnuts and extracted several tins of oil. I was sure that we could manage to live a healthy and sustainable life at the village.

After I started living at the farm, I changed my lifestyle completely. I switched over to using public transport for my travels. I took out the car rarely. Branded clothes were out and it was a long time before I bought any new clothes. Anyway at the farm, one was not expected to wear designer clothes.

A pair of shorts and a T-shirt were more than enough for the kind of work we did there. I realized that it was just a case of reducing the dependency on money.

The Water Diviner

One of the first things we had to do after we bought the land was to find drinking water since the river water was not potable and the village relied on wells. The nearest well for us was almost half a kilometre away. We decided to look for groundwater on the land so we could dig a borewell. People around the village relied on an old man who used a coconut to find water. This man walked around the field with a coconut in his palm. The spot where the coconut moved in his palm would have water or so he claimed. Everyone in the village seemed convinced that this would work and they claimed that he had a high success rate. Most of the water he found was around 30–50 feet deep.

We were not very convinced by the coconut method and decided to search on the Internet for a better option. Our search led us to Michael Davis, a water diviner. Water diviners are people who are gifted with the special power to find water even if it is deep underground. It is very similar to the coconut method except that they use special implements to assist them and are thought to be more reliable.

I went to meet Davis and was impressed by him and his work. He explained the difference between his search and a normal 'coconut' search at the village. His technique was to find geothermal water. This water is supposed to rise from the very core of the earth as steam till it reaches a sort of rocky aquifer and cools to form a stream of water under the rock. The coconut method is usually used to find water that is fed by the rains and is not reliable in seasons where the rainfall is less or nil.

He also gave us references and asked us to get in touch with people who had used his help to understand his method and success rate first-hand. We spoke to a number of people who had used his skills and were pleased that most of them were happy with the results. We thought Davis was a better option as he had a '98 per cent strike rate' and looking at the long term it was better to catch geothermal water than rain-fed water.

And so on a warm Sunday in May 2004 just before dawn I drove Michael to the farm. We had packed sandwiches and two huge plastic cans of iced water which he had specially requested. As soon as we reached the farm, he had a quick smoke. Then he walked around the farm to observe the place. He explained to me the importance of every plant and tree that he came across. As we passed the fig trees, he said that they were a pointer that water was always close at hand.

He pointed to the branches of trees like banyan and bamboo and said they always bent towards the source of the water. By walking around Michael made an assessment of the farm and used the trees to find out where water could be available under the ground. The clump of bamboo at the river edge of the farm had branches which sloped down to the ground. This indicated that water was near. The area with a few eucalyptus trees near the other edge would not be suitable since these trees usually sucked up all the water. He pointed to the beehives and said the bees too were an indicator that water was there in the area.

We came back to the house for sandwiches and lots of water. Michael then started off on his second round, this time with an old silver pocket watch which he used as a pendulum to dowse. The last time Michael talked a lot but this time he only muttered to himself. The pendulum moved to and fro and at certain points it moved crazily. At this point, he would stop and move the pendulum all around the area to ascertain where exactly it went crazy. I was reminded forcefully of Professor Calculus in the Tintin comics.

As he had predicted, near the fig trees the pendulum moved around furiously but it was at the bamboo clumps near the river that it moved the fastest. After the second time, he came back to the house to drink phenomenal quantities of water. When he went back to the field, he focused on the five points where he observed the pendulum moving really fast. He kept chanting to himself, watching the pendulum at these specific points. I noticed that he looked very tired and when he came back to the house, he took an hour's break during which he barely spoke and chanted under his breath. We had more sandwiches and vadas and after this Michael went back to the two spots he had selected, one under the bamboo clump and the other near the main gate.

This time he had additional equipment—two right-angled metal rods which he held in his hands with the angles away from each other. As he reached these two spots, the metal rods would swing towards each other and start vibrating. The other equipment he used was a forked stick, to the ends of which he tied a yellow plastic bag filled with water. The pendulum was now in his pocket. As he moved the forked stick towards the spot under the bamboo clump, the plastic bag started spinning around in circles. I saw the heavy water bag whirring around, and once or twice, the spinning became so furious that the stick and the bag flew out of his hands. When this happened, he asked me to move aside as he had no control over the movements of the bag. Michael would have to walk a few feet to collect the bag and the stick.

This furious activity went on for a while. In the silence, I could hear the 'whoosh' of the bag as it spun. It was fascinating to watch the bag go round and round crazily when he brought it close to the spot. Just to make sure I had a try at it and was surprised when I could not even make it go round once. It had to be some special power that made it whirr so crazily. I was thrilled and satisfied that I had the skills of a special person to find water.

Michael took out his pendulum and sat on his haunches on the ground to identify the exact spot where water could be found underground. He kept moving around the spot in circles with the pendulum to find the exact place to dig. This took about forty-five minutes, at the end of which he placed an iron rod on the ground.

I assumed this was the place I would have to dig for water. It was almost evening by then. He identified an alternative spot near the gate using the same rods and the plastic bag. Here too he marked the spot with an iron rod. Once this work was done, he broke his silence. He said the first spot was an ideal location for water but in case that was not possible, the one near the gate would be just as suitable.

He also gave us detailed descriptions of what would happen when we started digging the bore well. It was almost a month later while digging the well that we observed Michael's predictions coming true. Even the colour of the soil being thrown up was exactly what he had told us. Needless to say we struck water at thirty feet but went on to dig to 250 feet as per his instructions. This was to ensure that we had water even during the lean months.

I got the water tested at a laboratory in Mumbai to check if it was potable. Luckily for us the report was positive. This was one of the major risks that we had noted in our original plan. If there was no drinking water, it would have become difficult to continue living there.

6

First Year at the Farm

Groundnut Harvest

As the days passed by, and I spent more time at the farm, I learnt the finer points of growing various crops, their harvesting and care. I observed that each crop had its own method of sowing and harvesting. The pests and diseases that attacked each crop were also different. The treatment for each of these problems was a challenge in itself. To add to my problems there was no clear way to handle these problems organically. It was more of a trial-and-error method that gave results.

It was in November 2004 after the monsoon crop that we decided to plant groundnuts. Providentially, the day the groundnut seeds were being distributed I did not have enough money and could buy only two bags or 60 kilograms of seeds. The saga of the nuts started when we were informed gleefully by the villagers that we had to shell the groundnuts before sowing them. Well, we started in earnest only to find that at the end of the week, the two of us had just managed to shell only a quarter of a sack. This was the one time we really wished we had a large family with lots of uncles, aunts, kids and grandparents like most people in the village.

For that's what was going on in each house. The whole family used to spend the entire day shelling groundnuts. Even when you had visitors at your place they just sat with you and casually started shelling the nuts. We did have our fair share of visitors from the village but that was not enough.

Necessity, or in this case lethargy, is the mother of invention and by the end of the week both of us had decided that under no circumstance were we going to finish this task and we would plant the groundnuts as they are with the shell. Our argument was that if peanuts were intended to be without the shell nature would have done so. So that was it. We were breaking the norm and doing a crazy thing.

Word had spread that we were planting whole groundnuts and almost the entire village trooped in to tell us that we were doing something really stupid. Only a few old people in the village like Mohan and Baban's father said that many years ago they used to do the same thing.

So with a few rumours to help us and an article we read in *Honeybee* which said that some other farmer had also tried it, we went ahead and sowed the whole nuts. We soaked the seeds in water the night before to soften the shell. The last seed was sown on 31 December and we returned to Mumbai to celebrate the New Year, praying that the crop would not be a disaster.

A week went by with no action on the field and on the eighth day we saw a few saplings peeping out of the soil. Soon the entire field was covered with young groundnuts. We were whooping with joy till the villagers told us that a few shoots did not mean a thing. So much for encouragement. We watered and cared for the rows of plants, weeding, cleaning and tending to them lovingly.

Four months went by and in April when we pulled a few plants out we saw tender groundnuts hanging from the roots. In May it was harvest time. On 9 May we started pulling each plant out and leaving them to dry in the sun for a day. Meena took a week off from work to help with the harvest since there weren't too many people from the village. It all looked fine till we realized that the way to remove the nuts from the plant was to pick up a few of the uprooted plants and whack them on an iron rod tied between two posts till the nuts fell off. Sounds simple and easy till you realize that you have to do this from 7 a.m. to 6 p.m. and suddenly, we did not feel quite enthusiastic about our bountiful harvest.

It seemed a bit hard till we realized that it was an excellent way to vent our anger. So you pick up one bunch, think of someone you want to hit and start whacking them. A few bunches later you have a big smile on your face and the world is a much better place. By the end of the week (or should I say weak) we had finished only half the area we had planted and

Meena had to extend her leave by one more week. We were completely exhausted, sunburnt and covered with heat boils. Now, we were really glad we had planted only two sacks and not three.

After a marathon two weeks of whacking, drying and cleaning, we had filled eighteen sacks of groundnuts. That was way ahead of the village standard of 10–12 sacks for the same amount of nuts sown. News of this bumper crop spread like wild fire and we soon had people swarming in, this time to check if it was true. Lahu Kaka and Sridhar Kaka, my neighbours, were extremely encouraging and promised to check out this method of planting on their field the following season. There were a few sceptics like Baburao and Kate who thought it was beginner's luck but most agreed that it had worked.

Once this was done with, the next task was to lug all the sacks to the river and get the nuts washed to remove the mud sticking to them and dried again. At the end of the exercise we had harvested a whopping 707 kilograms of groundnuts waiting to be converted into oil. We had completed our first groundnut harvest and we were happy to see the end of it.

We got the oil extracted at the local oil unit. This oil was then double-filtered and packed into 15 kilo tins. We had thirteen such tins with us. We decided that there was no point in going to the retail merchant who would shortchange us anyway. We sent messages to our friends asking if they wished to buy the oil from us. The oil we had was 100 per cent pure, unrefined and unadulterated.

Since I was selling the groundnut oil I decided to pay a visit to the nearest oil manufacturer and find out how they did the refining process. This I thought would assist me in selling my oil which was only double-filtered. I was also keen on knowing how they managed to keep the price of oil so low. What I gathered from my visit was quite disturbing. The refining process was an automated system where they passed hydrogen gas through the oil. This removed the odour and also increased the shelf life of the oil. This benefited the manufacturer but had no intrinsic benefit to the end consumer. In fact, research has proved that the passing of hydrogen molecules during the refining process actually increased the chances of fat accumulating in the arteries which in turn led to other cardiac problems.

Though the person I met did not admit it in so many words he did indicate that some of the oil is adulterated with cheap and odourless oils like cottonseed oil. Another cheap oil used for adulteration was palm olein, which was diverted from the Public Distribution System into the grey market. This was how they managed to reduce the price and keep it low for consumers.

I was no scientist or doctor to comment on the accuracy of the information I had gathered. Anyway I did not need any additional information to sell the oil I had. Once people had tasted the oil and enjoyed its consistency they eventually ordered for more. Within a short period we managed to sell off all the oil we had produced. For the first time since we had bought the farm, we managed to recover the cost of production and also make a small profit.

First Anniversary

On 31 December 2004, a year after I left IBM to start this new venture, I did a stocktaking. That meant 365 days of not going to office, no PowerPoint presentations, no formal clothes and no shoes (except for a couple of weddings). The last one year was governed by running around for seeds, fretting over rain or the lack of it, numerous power cuts, dealing with corrupt officials, quacks and other scum of the universe.

It was an eventful year with many pleasant experiences (when we got the first crop of moong and struck water on the land), a few tense times (when the paperwork was stuck) and some traumatic moments (when we lost the entire crop of *tur* (pigeon pea). Overall, considering I was a novice at farming, the year was good though we did have start-up problems.

I couldn't resist the tradition in IBM and found myself doing a self-evaluation as the year ended. Unfortunately, here there is no boss to rate you or to argue with over the ratings. It is the truest form of self-appraisal with no space for skew.

After a good harvest of moong, we had a failed crop of tur and *urad* and a low yield of rice. But the vegetable patch was flourishing and the best news was that when we came to stay at the farm, often we did not carry any vegetable from Mumbai. There was plenty of choice too—from brinjals, carrots, radish, methi (fenugreek) and ladyfinger to cowpea and even string beans.

The infrastructure at the house was in place and we had a good toilet and piped water. A simple but well-equipped kitchen ensured that no one went hungry when at the farm. Our home at the farm was all set.

I felt more at home in the village and actively contributed to all local activities like Janmashtami and Navaratri and even sponsored the first prize for a local cricket tournament. I also tried to take part in the village meetings whenever possible.

During the last one year I realized that there were few people in the village who wished to stay back. Most of them wished to migrate to a big city or town and make some money. It did not matter what they did but they wished to move away from agriculture. Here I was trying to find my foothold in the village and eke out a living from agriculture while the others wanted to run away from it all. All the things that fascinated me like the river, the trees, the crops and the weather seemed unimportant to the villagers. A few of them felt I was on a sort of sabbatical, just having some fun and games at the village. They expected me to rush back to the city in a year or two when I ran out of money.

At the end of one year, though we did earn a little money, the fact that we could eat our own rice and vegetables was in itself extremely satisfying. Besides, the fact that we were now more welcome in the village and were being treated as one of them was a major triumph.

In this romantic and wonderful setting there was one thing in our original plan that did not happen. The job offer that Meena got on the day I resigned materialized and in April 2004, Meena joined *The Hindu* in Mumbai as a regular employee.

I had to contend living in the village alone though I did return to the city every weekend. We still did not have a telephone connection and the mobile connection was erratic. I also found it more and more difficult to keep transiting between the village and the city. Each time I went back to the city I would get a cold and cough which miraculously disappeared when I was back at the village.

It was clear that my system was not able to take these sudden bursts of pollution every week. Meena would try and come to the farm and spend more time but with her work and odd hours, it was not an easy task. While I was grappling with my peculiar situation, she too was not having a great time. She was alone in the city and finding it difficult. It was obvious that she could not go out with friends every day. We had to work to get to a solution to this problem at the earliest. Meena was doing well at work and she felt it would be too early for her to take the plunge and move to the farm. Besides, there was a monthly salary cheque which was welcome. It meant less pressure on us and more time to experiment and aim at living off the farm. Our initial three-year period had changed now. We decided that we would continue with the current arrangement till the moment was ripe for her to quit too.

I made it a point not to use any of Meena's income for the farm activities. Her income was used only for expenses incurred in Mumbai. The house in Mumbai was running on her income. If I used any of her income for expenses at the farm, it would defeat the purpose of the farm and the transition I was trying to make. I had a separate account for the farm and used the money from there for all farm-related activities. If I ran out of money I just postponed or managed with some work around till money was generated. The main source of income at the farm was the groundnut oil. The income we got from it just about covered the expenses of the farm. It meant that we were not generating loads of cash from the farm. Yet I was sure that we could live off the farm with the savings we had to assist us in case of any major expenses.

7

The Search for Rice

The Scent of Rice

The first year I was late for the rice-sowing season and had to resort to growing the GR4 variety that was short term and recommended by the agricultural officers at Kosbad. The next year we decided that we would start early and try to find some good traditional variety of rice to grow. We had read about traditional varieties of rice and knew that they did not require very high inputs of fertilizers. These varieties were also quite strong and resisted pests. We were sure that it was this type of rice that would grow well in our farm where we did not use any chemicals at all. Our previous year's experience and low yield had taught us a lesson and we were sure we would not plant hybrids this year.

In April 2005, we started to look for a good variety of traditional rice. It was one of our neighbours in the village, a businessman from Mumbai who owned land, who suggested that we plant a local scented variety of rice. Most of the farmers in and around the village of Peth had switched over to hybrids. The younger generation of farmers thought I was crazy to ask for the 'desi' variety, as they called it. My regular visits to the villages around searching for a good traditional variety also did not yield any results and we were almost giving up hope.

I decided to give it one last try and spoke to Baban's father and some other elders, that is, when I could make any sense of what they were saying. Most of them are too old to work and are drunk all day. In fact, they get pension from the government which, according to them, is meant solely for their alcohol consumption. A wonderful use for the pension scheme! Anyway, after many meaningful conversations, they mentioned the name of Kasbai.

Kasbai is a traditional long-grained rice variety which has a distinct aroma, though much milder than basmati. It's a long-duration crop and most of the older people remembered growing it years ago. But they all shook their heads when I asked them about the seeds and told me that it had 'disappeared'.

The tales of Kasbai made us more determined to get it. We decided that if we did manage to get some seeds this would be a great rice to grow. I thought the government may know something about. A visit to the agricultural officer was enlightening. He had not even heard of this rice variety. He said the villagers were taking me for a ride and there was no rice by this name. He rattled off the names of a number of latest hybrids and even offered to give me some of them free

The Search for Rice

of cost for a trial. Cursing myself for wasting time with him I moved on to the next destination.

This time it was the Adivasi Mahamandal at Kasa which buys rice from the Adivasi villagers on behalf of the government. Kasbai did not figure in their files. A good indication why people did not grow it any more. The market itself did not recognize the rice, so if you grew it you would not be able to sell it. However, the officer incharge here had more knowledge of rice and did remember Kasbai being sold to him a few years ago.

A few cups of tea and some gentle prodding revealed that the rice was grown four or five years ago in a nearby village called Dhanivari. Baban and I decided to go there. We were in for a shock as it turned out to be a sprawling village with hamlets scattered all around. We had to go back to the Mahamandal and request the officer for a name or a lead in the village. Looking into his records we narrowed our search to a farmer called Devu Handa. He had been the largest seller of rice to the Mahamandal last year and the officer assured us that he was a nice man who could help us out.

Back in Dhanivari, Baban and I started looking for Devu Handa and found a greying old man wearing a cap, sitting outside his house on a charpoy. An ex-sarpanch of the village, he had acres of land, a huge house and a large family. After exchanging the usual pleasantries we came to the topic of Kasbai. The mere mention of Kasbai and Devu Handa drifted into the past. His eyes turned dreamy and with a tremble in his voice he told us how the entire village at one time grew only Kasbai. He said, 'There was a time when people passing our village during lunchtime would be forced to stop and ask for a meal. Such was the alluring aroma of Kasbai.' The entire area would have this heady aroma hanging in the air as all the houses cooked the same rice. Today, he said, no one grew Kasbai and everyone had shifted to growing the new hybrid varieties. He claimed he had to force himself to eat this rice that was so insipid!

We asked him the reason for this shift and without a moment's hesitation he said it was all due to irrigation. He said that years ago there was no canal system in the village and they depended on the monsoon. With the advent of irrigation, farmers were tempted to grow a second crop and Kasbai, being a long-duration rice, was replaced by the shorter duration hybrids so that the harvest could be done earlier. This ensured that the farmers could take up a second crop.

I asked why he had shifted if he was so unhappy with the hybrids. No one forced him to, did they? He smiled and replied that their fields did not have fences and once the harvest was over the cattle were released into the fields. 'If my field alone has Kasbai it will be a treat for the cattle,' he explained.

'Sometimes, we have to fall in line with the community,' he lamented. Hybrids need more water, fertilizers and pesticides. He said that yields were good initially but of late, had reduced a lot. Besides he said that each year they had to increase the quantity of urea and pesticides they used. It was as if the newer hybrids had an insatiable appetite for chemicals. He told us that even when there were flash floods in the sixties, Kasbai had stood its ground. He fondly remembered how the rice was still standing when they all returned to the village after the floods had receded. 'Such was the strength of the rice. But look what we have done,' he rued.

As he went on reminiscing about the rice, we gently guided him back to the reason for our visit, the Kasbai seeds. He was sure that there was not a single villager in his area who would have the seeds of Kasbai. According to him, the only people who still grew it were the Adivasis in a hamlet at the foothills of the mountains in the next village Asarvari. We bid farewell to Devu Handa who lovingly blessed us and said, 'Mahalaxmi, the local goddess, will give you the seeds of Kasbai.'

In Asarvari village, we asked the sarpanch to help us as we were not very fluent with the local dialect. He sent his assistant Jeevan with us into the hills. After a half-hour walk through thick vegetation, crossing numerous streams and ditches and scrambling over rocks and gravel, we reached the sleepy hamlet of Boripada. There were just two ramshackle houses in front of us and we wondered if this was the right place. A wrinkled old woman sitting before one of the houses looked at us with curiosity. As we approached her we signalled to Jeevan to ask the crucial question. She muttered in reply and we looked at Jeevan for a quick interpretation. He broke into a smile and informed us that she did have the rice and wanted to know who we were and why we wanted it.

It was a difficult task to keep a straight face and I had to control a strong desire to hug her. After searching for months, we had found the elusive Kasbai. We explained to her that we were from Peth nearby and we needed the seeds to grow it. We asked for 10 kilograms of rice. She muttered and scowled. Jeevan interpreted that she had never heard of Peth village and also did not have a weighing scale. She was willing to give the seeds only in baskets. We asked for a single basket of rice and Jeevan told us to pay her something. I handed over a 100-rupee note and for the first time in the last ten minutes, her face broke into a smile. She nodded her head in approval.

As we walked back, against the fading sunset, leaving behind a smiling old lady, I couldn't help but wonder that here, nestling in the foothills of an unknown mountain away from the hustle and bustle of the road or the city, were the real people of India. These were the people who still held on to the rich biodiversity of our land and no one even cared about them. They had never heard of hybrids, fertilizers or pesticides. They just grew their rice and ate what they got. The old lady we met had probably never left Boripada. Her world was unspoilt by 'progress'. And for once I was grateful for that.

System of Rice Intensification

There is a lot being said about the System of Rice Intensification (SRI), which is an innovative method to grow rice. This method was initially discovered in Madagascar where one of the farmers noticed that even if the rice field was not flooded with water, the crop was excellent and the yields high. The method had then been adopted in India and was released for use by farmers by the Central Rice Research Institute in Hyderabad.

I was keen on trying this method and had to make slight modifications as I had no intention of transplanting the rice. I spoke with a couple of scientists and they felt the modified method could work, but they had not heard of anyone trying it.

The rice seeds had to be planted at a distance of 25 centimetres from each other and this was not possible manually. I devised a small rake using a plank and strips of wood. The wooden teeth were nailed to the plank at even intervals of 25 centimetres each. Baban and I then pulled this plank across the field, first horizontally and then vertically, to form a sort of grid on the soil. We got some help from the village and asked them to plant 4–5 seeds at the intersection of each square grid. It worked beautifully and within a couple of days we had sowed the entire field. All this was done before the rains.

The next week it rained and soon enough we saw tiny rice saplings neatly growing at a distance of 25 centimetres each. The rains continued without respite for a whole month. Along with the rains came another disaster. Baban fell ill and had to be hospitalized. With excess water and no weeding the grass grew as high as the rice and in some places managed to outgrow the rice.

Kasbai is a long-duration crop (160 days) and our land is on a height. The monsoon retreated after 120 days and the land started drying up. I could not install the pump to water the rice. By the end of the season we knew we had lost the entire rice crop. A couple of weeding sessions and watering towards the end would have saved the day but I was unable to do this alone. It was a tragedy that we could not save such an excellent variety of rice. I kept ruing the fact that I had failed to grow this elusive Kasbai.

I realized how modern agricultural methods with the so-called tested hybrids had nearly wiped out this traditional variety of rice that had once ruled these areas. The scientists at the rice institutes are promoting hybrids and now we are threatened with genetically modified rice. All this is expensive to grow and also does not have the same taste as the old traditional varieties. It would be a tragedy if all the traditional varieties disappeared and we were left eating insipid hybrid varieties which we are in many cases.

Years later, when my friend Vipul bought land, he managed to grow Kasbai and save the seeds too. He still grows the rice every year and we buy it from him.

8

The Present Scenario

It was more than nine years after we bought the land and had planned to move to the village that the move finally happened. Meena was doing well in her job and was soon promoted to chief of bureau, *The Hindu*, Mumbai. This meant longer hours and more responsibilities. Even when she came to the farm to stay she would be monitoring the news and coordinating from here to ensure that the reportage for the paper did not suffer. It was in 2013 that she was offered a posting in Pakistan as the paper's correspondent there. Once she left the country, I too packed my bags and moved to the village.

The last nine years, I stayed at the farm during the week, returning to the city for the weekend except during harvest or sowing when I stayed back to finish the work. I felt like I was on an IBM project where I used to shuttle between our project site and the city. During the nine years at the farm, there was a lot of learning and unlearning that happened.

I experimented with a lot of crops to see what all I could grow on the land. In my quest to grow our own food, I experimented with turmeric, mustard, wheat, jowar, maize, varieties of pulses and oil seeds till I found which ones were ideal to grow on our land. One of the first lessons I learnt was that everything does not grow everywhere. Each crop was dependant on the soil conditions, the temperature, the humidity in the air and the water content. What did well in one area will not necessarily grow in ours. Also the selection process is a slow activity. When you experimented with one variety and found it did not work well, you had to wait for the next season to try the next variety or to rectify the mistakes you had made in the earlier season.

Besides the crops I sowed each season, I also learnt a lot about the fruit trees we planted as the years went by. I planted a couple of cashew trees which gave ample fruits every year. The Alphonso mango did not grow well as compared to the Kesar or Malgoba variety which flourished. This I learnt only after five years when they started fruiting. I replaced all my Alphonso trees with the Kesar variety, but I had lost five years before I did the switch. I planted passion fruit which grew well at the farm and the drink we make from it is very popular.

In the city, when one shops, you can only see the produce. No one knows when it was grown. There are different varieties for different seasons. The moong you grow in the monsoon is the Kopargaon variety while the one you grow in rabi is the Vishaki variety. Both look the same after the harvest. I learnt this the hard way by sowing the wrong moong one season and realizing the mistake only when the crop failed. It is the same for all crops and one has to be careful when selecting the seeds.

After our first two crops of rice had been a failure, we tried and planted as many traditional varieties of rice as we could find in the area. After a few seasons we finally decided that the best rice in terms of taste and yield on our land was the Kala Karjat. 'Kala' means black and 'Karjat' was the variety. The husk of the rice is black, which is from where it gets its name, and the aroma and taste suited our palate. We decided that we would continue with this rice for our consumption till we find a better one, which we have not till date. We take the rice to the local rice mill where I pay him extra money to give me unpolished rice. It is amusing for the mill workers that I do not get the rice polished and they all crowd around and hold the rice and smell the same. Once one of them even remarked, 'Seth comes from the city and eats unpolished rice, while we from the village have stopped eating it.'

Besides Kala Karjat we also heard of a red variety of rice that the local villagers had used years ago, but long abandoned for the newer hybrid varieties. I looked all around for the seeds till I heard of an Adivasi in the nearby village who had and was still growing the rice. I contacted my friend Raghu Valvi from the village and asked him to try and get some seeds for me from this Adivasi. Raghu managed to get me a few kilos and we have been planting the same rice since then. It is called Kudai and after dehusking the rice is dark red in colour. Meena likes to eat the rice cooked plain though I am not very fond of it in that form. We use the rice to make dosas, idlis and other dishes which need ground rice.

The villagers were excited when they heard that I was growing Kudai and some of the elders even shook my hand. They recollected how they also used to grow the rice and eat bhakris made of it. They told me, 'The bhakris made from the new hybrid rice tastes like paper, but the bhakri from Kudai is like nectar.'

Besides selecting the correct variety, one had to also experiment with the method of sowing. All around us the method used for rice plantation was the transplantation method. Nurseries are prepared at the start of the monsoon and after a month the rice is transplanted into the fields. The entire family and sometimes even labour from the nearby villages were employed in transplanting the rice. I was sure I would not be able to handle it. Besides Baban had his own field to transplant and there was no way he would come to work at my place. I tried and tested many ways of growing rice with less labour and finally worked out a way to do it. I waited for the rains to fill the field and then ploughed the land to make it muddy and sticky. This way the grass was buried under the mud and reduced the effort of weeding. I then just broadcast the rice and let it overcome the grass. It worked well for me though the yield was low and did not match the village standards. The rice we got was enough to feed us with a little surplus to be sold off in Mumbai.

The Present Scenario

As the years went by, I experimented with some new crops each season. At first I tried everything on a small scale and then when I had a fair knowledge of the crop cycle and its finer details I increased the area. We now grow sesame and groundnuts for oil, two varieties of rice and a range of pulses like tur, moong, urad and *val* (field beans). Besides, each season we plant different vegetables that grow well in our area. Ladyfinger, pumpkin, brinjal, cluster beans, string beans, carrot, radish, sponge gourd and yam are some of the vegetables that do well on our land. I now have pepper, mustard, turmeric, basil, lemongrass, allspice tree and ginger as well. The turmeric from our land is pure and even when just one pinch is added to the food the aroma fills the entire house. It only reconfirms our belief that there is rampant adulteration in the market.

The betel leaf plant we got from Kerala is very popular in the village and they all come to take a few for their religious ceremonies all through the year.

I managed to get in touch with the Coconut Board of India which has an office in Palghar and got fifty coconuts which I planted all over the farm. They are just two years old and though a few did not survive the rest are doing well.

The mangoes, gooseberry and lemon grown on our farm are made into pickles every year which easily lasts us through the year. The mango pickles range from the baby mango pickle to the large cut mango pickle and the shredded mango pickle.

The variety of crops increased and now I was worrying about crop yields, the rains, composting, seeds and the like. Looking back, I had come a long way from that evening when in a moment of madness I decided to quit my job at IBM without even owning a plot of land.

In 2014, Meena was expelled from Pakistan and posted to Delhi. After nine months she was transferred to Chennai and that's when she decided enough was enough. She finally decided the time had come for her to quit and join me at the farm. Her job had made us financially more comfortable as compared to when we started out in 2004. We decided to dig into our capital and make some essential changes to our house. The asbestos roof had cracked in places and was leaking during the monsoon. Also we had a rodent problem. They were living between the gaps in the roof. Though they were harmless except for making irritating scampering noises all night, it was the predators that came after them that worried us. We had already spotted a couple of rat snakes sneak under the roof looking for a fresh kill. A pair of bronzeback snakes was residing under the roof in the front porch too. They are not poisonous but it was unnerving to sit in the front porch with a snake slithering over your head as it went about hunting geckos.

There is a lot one needs to learn and adjust when you shift from the city to the village. There are many things that we take for granted in the city which are non-existent in the village. The 24/7 power supply, the newspaper vendor, the milkman, the garbage collector, the excellent mobile network, the corner shop, the cobbler and the entire service industry which is at your beck and call is missing in the village. No one comes to do the daily household chores or small tasks in the village and one has to manage on one's own. The Do It Yourself (DIY) culture in the western world is the norm in the village. I soon got my own set of equipment like a drill machine, a spanner set, a hammer set and other carpentry tools for repairs at the farm.

Living in the village brought us closer to the realities of the rural economy. It is one thing to sit in the city and read about farmer distress and suicide and another to actually see it upfront. I used to always wonder why someone would end their life because they could not repay loans as low as Rs 20,000. Now I could understand the trauma and distress they went through when a crop failed and they had no money. If a money lender or official visited someone to recover dues, it wasn't long before the news spread though the village. One of our neighbours had their television set taken away by the credit recovery guys as they could not repay the loan they had taken during the drought in 2015. The lady of the house was so embarrassed with the episode that she did not leave her house for almost a year.

The price of crops and marketing was a major sore point with everyone. From the first season, we had problems marketing our produce. After our experience with the first crop of moong, we started retail sales. Over the years, we have tried various experiments with marketing. We even tried to set up a retail vegetable enterprise in Mumbai but had to give it up soon. It is not easy to produce and market at the same time. Our experience taught us that doing both was not doing justice to either.

The struggle to find labour to work on the land is a constant one. With salaries zooming in the manufacturing and service industry, the young are not interested in working on the land. The few old people have their own land to till and finding time to work at our farm was getting increasingly difficult for them. I scaled down my production so we could manage with the little labour that we could get and thought of ways and means of reducing the labour on the farm. From our initial two sacks of groundnuts, we scaled down to one. Large-scale vegetable farming was stopped and we planted only to meet our needs.

Living in the village also meant dealing with snakes and scorpions. There was no wildlife except for boars but they were in the mountain close by. While they did leave us alone most of the time, we had a couple of close encounters. While snakes usually steer clear, I was once doing some work near a chikoo tree when I accidentally disturbed a hornet's nest. It stung me on my upper lip and for a couple of days I walked around looking like an incarnation of the Hindu monkey god Hanuman, much to the amusement of the villagers.

A Day at the Farm

One of the first things that we realized when we started staying in the village was the complete lack of household help which we are so accustomed to in Mumbai and other cities. The women in the village, like everywhere else in the country, did all their work and more. They thought little of waking up really early, cooking, cleaning, washing clothes at the river, spending a full day in the field and returning to do the evening meal and other chores. I tried a lot to get someone to help me clean the house at least, but no one was willing to do it. At first, I thought it was because I was alone that no woman wanted to come for housework. After Meena moved to the village in 2015, we tried again but could not find anyone who was willing to come.

We divided the housework between ourselves. While she cleaned the house and made the rotis, I cooked the meal and washed the clothes. We bought a washing machine after a tiring year of washing clothes and consoled ourselves that it meant using the same amount of precious water.

I have always been an early riser from childhood and at the farm it was not different. I usually hear the cock crowing around 5.30 a.m. and get up, if my cat doesn't wake me up earlier, that is. In summer I can hear the familiar sounds of bulbuls, mynahs, crow pheasants and sometimes sunbirds and some others. I then start the coffee machine and cook chicken legs and rice in the pressure cooker for the pets. We have three cats now, Minimus, Whitey and Dada, and a dog, Pepper. Pepper will already be at the front door, waiting for me to step out. After the mandatory petting we both go and sit by the river. I have my coffee there. The opposite bank of the river teems with children from the

Adivasi Ashram Shala School, bathing and preparing to go to their classes. On some days we spot the odd fisherman working his nets in the river. While I sit and stare at the river, Pepper sits on her stone platform that we built for her next to the chapra.

A good half hour later both of us return to the house. By this time the cooker has cooled and the food is served to the pets. Pepper is fed in the front porch while the tomcats get their share in the back porch on a table. Mini the princess of course is served in the kitchen. She refuses to eat with the others.

I usually walk for forty-five minutes every day, sometimes Meena walks too, and this is also the time Pepper loves to play. The moment we start to walk she will grab her play bone or a stick and start running before us. She runs up and down with the bone, daring us to take it away from her, growling if we came close and taking off again. Rarely do we get a chance to steal the bone from her and she usually gets tired after a few furious jumps.

The walk gets over by 8 a.m. after which I start preparing our breakfast. It is usually idli or varieties of dosa made from our own rice and dal or poha (puffed rice) made from the red rice Kudai.

As breakfast is being prepared Baban usually arrives with chicken legs from Baburao's son Sagar's shop for the pets' next meal. He then ties Pepper to the front porch and releases the hens. I discuss the morning tasks with him and he goes into the field to start his work.

The Present Scenario

Breakfast is usually from 9 to 9.30 during which we also firm up the menu for lunch. After breakfast I join Baban in the field for the task of the day, while Meena starts with the cleaning of the house.

We usually have a tea break around 10.30 by which time the house cleaning would be done with. A quick cup of tea and Baban would resume his task. I stay back at the house to begin the preparations for lunch.

Our lunch usually comprises of rotis, dal and a vegetable. We have a variety of pulses at the farm and so we make a different dal every day. There is moong, tur, val, chawli and chana. For variety we eat rice and a south Indian curry like sambar or rasam on some days. Most days we have lunch by 12.30.

A brief siesta from 1 to 2 and it is time to go back to work. Baban arrives after his lunch and we herd the hens back to their cages. Pepper is then unleashed. Once released, she accompanies us all over the farm following our activities.

Around 3.30, we have a tea break, and then back to work. I break around 5.15 to return to the house to feed the pets. Dinner time for them is 5.30. Meena usually does her workouts around this time. At 6, on most days, I leave with Baban to go to the village. This is a quick drive to the tea shop to catch up on the day's gossip or visit some ailing villager.

On my return, most evenings we sit by the river till it is dark and the insects arrive. Back home by 7, dinner preparations begin. It is also the time to check our emails and see what is happening on social media. After a quick bath, we either read a book or watch a movie from the vast collection of DVDs that we have accumulated over the years. Meena does her writing during the day in between her household work and has a fairly well-established routine.

Dinner is usually around 8.30 p.m. I go to sleep by 9 p.m, while Meena sits up and reads, watches films or works on her computer till 10.30 p.m.

The schedule of the day gets a bit hectic when it is the sowing or the harvest season. Those days the helps arrive by 7.30–8 a.m. and stay till 11.30–12 noon. After lunch, they resume work from 2 to 6 in the evening. On those days, Meena cooks as well, so I can spend the day on the field.

The work at the farm ranges from pruning trees, weeding, watering and checking for pest attacks besides the usual sowing and harvesting. Before sowing, the land needs to be prepared with the grass cleared and the required preparations done which varies from crop to crop. There is a lot of planning that has to be done as each activity is time-bound and has to be done at the right time. Even a couple of days' delay in watering the crop can lead to a huge dip in production. With the erratic power supply, the planning process becomes even more complicated.

There are days when a trip to nearby Kasa would be required either to get some provisions and fish for the cats or to catch the power guys in case of a drop in the power supply. I have a three-phase motor to draw water and a drop in one line means that the pump will not start.

Recognition

A year after I had quit my job and started farming, the *Times of India*[1] published an article on my transition from a corporate professional to a farmer. It did evoke a lot of response at that time and we had a stream of visitors from the city coming to the farm to check what we were doing and also to understand how we were handling the transition. Many even expressed a desire to buy land and farm like we did, though I doubt if anyone actually got down to it.

A few more articles[2] on my life as a farmer appeared in publications like the *Hindustan Times*[3] and *Mint*.[4] We were even interviewed on the television channel *Times Now*[5] in a chat show. These articles and media coverage did bring in visitors and calls from people, mostly from the city, who appreciated what I was doing and were encouraged by my decision.

---

[1] Nisha Prabhakaran, 'Keyboard Charm to Organic Farm', *Times of India*, 3 January 2005, Education Plus.

[2] Lekha Menon, 'New and Improved', *Mumbai Mirror*, 16 June 2010 and Kunaal Majgaonkar, 'Farming Nirvana', *Times of India*, 19 May 2007, West Side Plus, vol. 7, issue no. 12.

[3] Naomi Canton, 'Out of IBM, in the Fields and Gazing at the Stars', *Hindustan Times*, 10 August 2007 and *Hindustan Times Cafe*, 'Pure Magic', 6 February 2009.

[4] Livemint, 'Open Sky Policy: Mumbai to Peth Village', 8 September 2007.

[5] *Times Now*, 'Life's Like That', 5 July 2005.

In 2011, a Marathi daily *Agrowon*,[6] a paper dedicated to farmers and farming activities, decided to feature me in their annual Diwali issue. The journalist Vijay Gaikwad paid a visit to the farm and spent a day understanding what I was doing. The detailed article along with pictures of the farm was published in the Diwali issue and what followed was totally unexpected and overwhelming.

I started getting calls from farmers from across the state congratulating me on my work and decision. For almost a month I took 25–30 calls per day talking to various farmers who said they felt encouraged after reading my story. They felt that if a person with a steady job in the city could move to a village to farm then they were surely in the right field and hoped they would be able to get over the current crisis that faced the agrarian community.

Initially I started noting down each caller's number and name till the calls were so many and so frequent that I had to simply abandon my jottings. I was happy that I had managed to generate some sort of confidence in the lives of so many farmers across the state.

There were some calls which were scary too. A young boy called from Bengaluru, saying that he was working for an IT company and hailed from a remote village in Maharashtra. He had got married recently and his wife was still in the village while he earned his living in Bengaluru. After reading my story he had decided to quit his job and

---

[6] Vijay Gaikwad, 'IT Toon Sheti', *Agrowon*, 2011, Diwali Issue.

go back to his village to start farming. I had to caution him and ask him to do his homework before taking any rash decision.

There was another call from a person from Nasik who was contemplating suicide since his crop had failed. He called to tell me that he had postponed his decision after reading my story. I spoke at length to him and promised to help him in any which way he needed.

There were other callers who just called to say how they were pleased with what I was doing and so on. Some of them would call at 5.30 in the morning and say, 'As a farmer you must be already in the field by now so we called early.' A few of them became so irritating that I started saving their numbers as 'Avoid Calls' on my phone and never answered them.

Years later, in 2016, the same story got circulated on WhatsApp. The reaction was similar to the one we had got after the print story. A lot of people called just to congratulate and appreciate what I was doing. Some people visited the farm to see a city slicker coping in the village. A few of them were keen on emulating me and grilled me on how to go about it.

I always shared my experiences with them and told them about the 5 Cs that they needed to have to do something different with their lives. They were:

- Courage
- Commitment

- Conviction
- Cooperation
- Capital

We had got an amazing response and I felt that I had indeed done the right thing by deciding to be a farmer.

9

Settling Down

By July 2004 we had finished setting up the house with all systems running. Many things still hadn't been done, like painting the house, but we had exhausted our budget.

It was not until 15 July 2004 that I actually spent a night at the farm with all the workers who had become my friends. They were scheduled to leave for Mumbai the next morning and we had a small party that night. After they left, I was on my own and it was a strange feeling. The first night alone at the farm is an experience I will never forget. It was completely dark since there is no ambient light like in the city. It was quiet except for the occasional cricket or gecko calling its mate. I left the porch light on to break the darkness. It was not of great help as the light cast eerie shadows keeping me awake. I would imagine weird noises and keep thinking that something had entered the house. That night I must have got up at least a dozen times to check the whole house for aliens.

In some ways I was reminded of our treks when we slept in strange caves and temples. But on the treks we were usually in groups which added to our security. Here I was all alone with the nearest human at least a good 200 metres away. I was sure that even if I screamed no one would hear me so far away and even if they did, I was not sure they would come to my assistance. Once I got used to the silence and the darkness, I found that the nights were extremely peaceful and I always got up early feeling absolutely refreshed.

There are no landlines in the village and the advent of the mobile phone was like a revolution. People kept calling each other for the sheer thrill of being able to communicate finally. For a while we did have the Tata wireless phones but they stopped the service some years ago.

The village has electricity though there are only five official meters in the village. There was a time a few years ago when the entire village had meters and official connections for power. But non-payment of bills resulted in the disconnection of power supply and the meters were taken away. It all started with one villager not paying the bill and stealing power. This crime caught on and soon enough everyone stopped paying the bill. Every evening people would step out of their houses with long poles with hooks at the end of wires to steal power. It is an extremely dangerous activity and Natwar, one of the villagers, was electrocuted in June 2004 while attempting to put the hook on a wet day. I thought that this would be a lesson to all of them. I was stunned when the very next day I saw Natwar's wife out with the familiar pole, doing the same dangerous thing that had taken her husband's life only a day earlier.

Over the years things have changed. The state electricity board has introduced stringent rules for people who steal power. There were rumours that they had taken away someone's television set when they caught him stealing power. The board also introduced a cheap power connection scheme where one had to pay only Rs 15 for a power connection. Within a couple of years every house in our village had a meter. It is only in the neighbouring Adivasi hamlet that there are still a few houses stealing power.

Our village is too far away for any newspaper vendor. The only way we get to read the paper is if someone brings it from Kasa, about 10 kilometres away. Kasa has a huge market, a hospital and two banks.

Initially, when I started staying in the village, every morning I would feel an urgent need to read the paper. I even went to Kasa to look for a vendor. One day one of the village elders asked me if there was anything in the paper that was actually relevant to the village. It struck me that he was right. Even if I read about some scam in the government or some bridge falling down in another country a couple of days late it was perfectly all right. Anyway, even if I got it hot off the press, what was I going to do about it? Nothing.

Things were to change in the years to come. Social media applications like WhatsApp became a source of information though the accuracy was sometimes suspect. With better Internet connectivity I could check the news on websites.

The need to read a paper in the morning, though strong, was manageable now. Besides everyone in the village except us had cable TV so they got the news as it broke.

What was important was the weather forecast, especially pre-monsoon. In spite of all the fantastic technology we had, the predictions were never right and even if they did come close to being correct, they were too broad-based to have an impact in our village. Meena used to send me messages on the monsoon bulletin. If monsoon showers were expected in Maharashtra in the next twenty-four hours, we would find that it reached the village a good six days later. They were not incorrect—it did rain in Maharashtra in the next twenty-four hours—but it is a huge state and the predictions were too wide-ranging to be of any use to the farmer. I found it better to rely on local knowledge. It was simple. If the mountain across the village had mist till its base it was expected to rain the next day. And rain it did.

While I got information on the monsoon from the official meteorological department website or from Meena's daily monsoon bulletin, Sitaram Kaka from the village relied on his almanac for the monsoon predictions. It is a game that we both played where I predicted based on my information and he based on his almanac. Days later we would compare our predictions and most times he won.

There was enough work every day to keep you busy. If there was no specific task, there was always weeding to do. The weeds grew as if they had some magic potion and no amount of weeding could stop their growth. I spoke to

Settling Down

some professors at the Kosbad Institute but they could only recommend some vile chemical which I had no intention of using. The other option they gave me was simple—just weed regularly.

Some days were hectic. If we had more work we would get help from the village. I knew most of the people in the village by name and was aware of which families were in need of money. I had built a reputation of paying on time and most of them were eager to work for me. Of course, they came provided it was not sowing or harvest time in their fields. We learnt to time our activities according to the village schedule so that we completed our work before they started theirs. We had the benefit of a river and a pump to draw water.

When there were many people to work it was resource planning and deployment that was important. I had to ensure that each group had the right mix of people and they did not end up fighting with each other. Each task could be done only by a certain number of people. Too many people could ruin things. Somewhere, I was glad of my project management training and the numerous projects I had completed for IBM.

To be able to do this one needed to have a good grasp of the village dynamics. Baban was my mentor and guide in this matter. How else would I know that Kamu did not speak with Parvati or that Nilesh's brother had tried to stab Sridhar Kaka years ago and relations were not cordial? It was a continuous effort to ensure that everyone worked in unison and went back home safe. They may quarrel outside the farm but I was not very keen on having a bloody riot on my premises.

When the villagers learnt that I was staying at the farm alone, I got a few visitors every evening. They would come in groups and sit on the porch and chat with me. I would politely offer them tea to which they would reply, 'Is this a time to drink tea?' I was unable to understand why they kept asking if they could leave after a while. I would reply in the positive and they would all leave immediately. Finally I asked Baban what was happening. He burst out laughing and explained the reason for their visits. The norm in the village was that in the evening if anyone came home, you offered them a drink. These guys who were visiting me were looking for a free drink. I quickly spread the word around that I had no intention of distributing alcohol and within a few days, these unsolicited visits stopped.

It was not a norm for people to stay alone here. I noticed that if due to any reason someone was compelled to spend the night alone they always invited a neighbour or friend to stay over. The fact that I was staying alone at night was a novelty for many in the village. There were also people like Baban's mother and sister who limped all the way to the farm to check who this person who stayed alone was. Besides, the fact that I cooked and cleaned myself without any woman to help me was fascinating for them. It was also unusual for the menfolk to stay away from their wives for so long.

The other thing they could not understand was why we did not have children. I got suggestions to visit the best doctor in the area who could resolve my problem so that we would be blessed with a child. Sridhar Kaka was so keen on getting us a child that he even got us an offer from a nearby village. A couple there who had three sons were willing to offer their youngest to us for adoption. Adoption was common in our area and they managed it quite well without any legal procedures and paperwork. One just needed five people from the village as witnesses for the adoption and it was sealed. We had politely refused these offers.

At first, it was the evenings that were difficult to pass. There was no television or radio. No Internet to surf and no one to even chat with. It is too dark to venture outside the house and even when the villagers did, they would be equipped with a stick and a torch. There were too many reptiles to take a chance in the darkness. Initially, after Baban and company left around 6 p.m., I found it very difficult to pass time till I crawled into bed. I had no idea what I could do. There was actually nothing to do. Yes, I could read but then that was possible only if there was electricity. It was much later in 2015 that we got a battery backup.

I thought about the city and wondered how many people in the city could actually spend an evening without doing anything. I realized that in the city, I was restless too and always looked for something to do. I could watch television or listen to music or the radio or go for a walk or meet my friends. I don't remember a single evening when we would sit around doing nothing in particular. As the days passed, I found peace in myself. There is a calming effect when one does not have to do anything in particular. It is almost akin to meditation. Something in the city made people restless; one had to do something to pass the time.

It was not long before that I learnt to pace myself. I cooked dinner or watched the stars or just sat in the darkness on the porch listening to various insects. A couple of years ago both Meena and I had attended a course on amateur astronomy and that came in handy on these lonely dark nights. I moved my dinner time to 8.30 p.m. so I could go to bed early. I was always an early riser and at the farm the chirping of the birds woke me up much before the first rays of the sun had shone.

Later when Meena moved to the farm and we kept pets, the schedule was more or less the same. We shared the housework and evenings were no longer lonely.

Breaking from the Trap

When I initially took over the farm, I was completely dependent on Moru Dada for my agricultural activities. It was he who got the gravel for the road, the electrician to check my pump and also arranged for Baban to help us. Everyone around seemed to be either related to him or indebted to him in some way. He was the master fixer; one just had to ask him and he would get things done. As I spent more time at the village I realized that it may not be a good idea for me to be so dependent on him. I also discovered that he had been overcharging me for all the services he gave us. It was not a large sum but he added a small commission for himself in each transaction. It had not taken me long to realize that I was in his vicious trap and conclude that I had to get out of it.

Moru Dada was sure that I was here only for a short while and would soon be running back to the city to make my millions. He used to taunt me every time we met by asking me if I was missing my 'computer job', as he called it. He was using my short tenure at the village to make as much money as possible from me. I knew what he was up to but kept postponing the eventual showdown.

The last straw was during the first monsoon season. I kept asking Baban to arrange for a plough and a pair of bulls to till the land. His replies were elusive and he kept hinting that we should hire a tractor. He kept telling me that there was no one in the village who had the time to hire out their bulls and how the tractor was best for the kind of soil we had at the farm. I was not ready for this since that meant a huge expense which I did not wish to incur. Besides, the entire village ploughed the land using bulls and the traditional plough. Why should I be the one to use the expensive tractor? I had also read that the tractor was not good for the soil as it dug too deep and turned the soil too much.

It took me a week and a lot of discussions with the other villagers to realize what the game was. It was Moru Dada who had instructed Baban to ensure that I used the tractor in my field for the monsoon season. I learnt that Moru Dada had helped Baban financially a few years ago and he was indebted to him. Moru Dada was the only one who had a tractor in the village. He used to make a neat packet every time he rented out his tractor. There were not many around who had the resources to hire a tractor and he was keen that I hire it from him.

I knew that I had to break out of his grip at the earliest or I would be bankrupt pretty soon and heading back to the city. I went to the village alone one evening and visited each house asking for bulls and a plough on hire. It was not long before I had found Jairam Kaka who agreed to do the work for me. I learnt that Jairam Kaka had three marriageable daughters and was always looking for work. Renting out the bulls and the plough to me meant a lot of money for him. I fixed the rates with him and he promised to turn up the next morning. I decided to be frank with him and told him about Moru's interest in this activity. Jairam Kaka was clear and firm when he replied, 'I do not work for Moru nor do I have anything to do with him. I shall be there tomorrow morning.'

The next morning Baban was shocked when he saw Jairam Kaka with the plough and the bulls ready to till the land. I took this opportunity to explain to Baban that I did not have much money and this was the only way I wished to till the land. He realized his game was up and confessed that he had been told by Moru Dada not to engage any help from the village but to convince me to hire his tractor.

I knew that this development would not be well received by Moru Dada and so I called Raajen the next day and asked for his help in convincing Moru Dada that I did not need his assistance at the farm. The call did the trick and since that day Moru Dada never interfered with my farm activities. We never spoke of the incident and were still very cordial to each other. If I needed any advice I still went to his house and he was more than helpful. Only this time around it was information that he gave me and we had no financial transactions together.

Village Community

The village is a huge community. Everyone is related in some way or the other. A single death in the village and almost half the men in the village would have shaved off their hair. Every family tilled their own land and only a few in the village could afford to hire outside help. Other than agricultural activity, the other major events were weddings and religious festivals. Both were celebrated with pomp and splendour and everyone was part of the celebrations.

Since I stayed in the village, I was invited to these weddings too. I usually attended them and though they were initially very different from city functions, over the years, buffet dinners, event managers and DJs have brought them up to scratch. The women of the village cooked the food at the weddings and the village boys usually did the serving themselves. I too took part in the serving of the food at the weddings.

I could not help but compare this with living in our flat in Goregaon, where we hardly met our neighbours and sometimes did not meet or speak with them for months together. We were all so busy with our jobs that we had no time to meet except during the usual bimonthly society meetings where it was more of a slanging match.

The village community brought back memories of my childhood in the railway colony where we all lived together as one family. Every festival in the Hindu calendar is celebrated with gusto at the village. For Janmashtami, little pots of milk and curd or *dahi handi* are strung up at varying heights for human pyramids to break just like Lord Krishna is supposed to have done in his childhood. The money attached to one such dahi handi is a princely sum of Rs 165, a far cry from the lakhs at stake in Mumbai. Gangs of boys or men form a circle under the handi. They sing a few traditional bhajans and songs, which warn everyone that they are going to break the handi and steal the curd from it. At this point the women of the village gather around and throw large quantities of water in a bid to discourage them.

The teams don't give up and form small human pyramids with the youngest person climbing right to the top. He then proceeds to break the pot and take the loot from inside. The pots are filled with milk, curd and a lot of fruits. As the loot falls down all the children rush to gather the fruits and the money that has spilled down while the pyramid collapses in a huge tumble. The first time I saw this, I was pleasantly surprised at the simple and orderly manner in which they had enacted the legend of Lord Krishna. It was so unlike in the city where we had blaring film music and shoving and pushing as people tried to break the handi.

Diwali is extremely quiet compared to the noisy and polluting one at Mumbai. There is no money out here for them to burn. Each family made sweets and distributed them to the guests. Again, I was reminded of the Diwali we had at the railway colony with tons of sweets and fun.

Navaratri is one occasion where the women and the men dance together. I go for these evening dances but usually return home early as in the night drinks flow like water and most guys get pretty drunk. Not that they did any harm but they would all cosy up to me and give great advice about the farm and make rash promises of help.

The Ganapati festival is like a trip down memory lane. The idol is not made from clay but bought from the nearby shop. This reminded me of the lovely idol my mother used to make of clay and the gala time we had immersing it.

Over time, I have built a good rapport with the children in the village and encourage their cricket tournaments and even accompany them to matches. Every village has a tournament and these limited-overs matches are played over two days between teams of eight players each. Our village too organizes a tournament for which we sponsored the first prize one year. It was extremely touching when I was invited on stage at the inauguration and honoured with a shawl for doing so. Till date our village has not won the first prize but we have not lost hope.

The summer holidays are fun, too. All the kids who are away studying would return to the village. It is also the time when most of the trees bear fruit. There would be plenty of jamuns, black berries and dates at the farm. The kids would come to the farm in the evenings and we would leave in a big group looting all the fruit trees. On those days I had to skip dinner as I would be too full with just the fruits. I went back to my childhood during those days and would eat along with these young kids without any embarrassment.

The other exciting activity is fishing in the river. Sometimes we fished with nets or with fishing rods. We hardly caught anything but it was the activity that was enchanting. Just sitting by the beautiful green water with a rod in your hand, waiting for a bite, is a feeling difficult to describe. There is hope, joy and a feeling of relaxation. In case we did catch something, I would give it away since I don't eat fish. The riverbank has scores of dragonflies of various colours. Once, I tried to catch a blue one to test my hand-eye coordination. After a couple of attempts, I had a blue one in my palm. I was thrilled that I was still quick. Of course, there was no Rex or Krishna, my childhood friends, to impress. I just let it go.

Shh . . . The Gods Are Coming

Like most villages in India our village too has a Gaon Dev or a village deity. Under the huge banyan tree in the village full of beehives is located a three-foot-high statue of the Gaon Dev. It has a human face with big eyes and ears. The earlier statue was made of wood and had started to rot. The local carpenter cum sculptor Baban Thakur made the new one out of stone. The stone deity is covered with vermillion or kumkum.

The deity is worshipped twice a year. Once before the rice is sown and once after the rice is harvested. Money is collected from each house and a couple of goats are bought. In the evening the menfolk gather around the deity and start howling. This is supposed to bring the deity to life. Then the goats are sacrificed amidst more howling. A fire is lit and the entire goat is barbecued. The smoked goat is then apportioned to each family which has contributed to the festivities. Depending on the size of the goat one usually got about 200–250 grams of mutton.

I did get my share of the offering but could not eat it. The smell of burnt flesh was not exactly appealing to me. Until recently I would send my share back with Baban. After we got Pepper, my share went to her and she just loved it. Besides the biannual Gaon Dev celebrations there is a ritual performed every ten years to cleanse the village of evil spirits.

Decades ago, on the eve of the first new moon after the monsoon harvest, the over-six-foot-tall head of our village, Rama Thakur, would patrol the narrow lanes, looking into each house after sunset. This was to see if any womenfolk or roosters or cattle had been left behind. Any violation would result in a fine and confiscation of the animal.

Thakur has passed on but there is a diligence in observing the Gaon Dev ritual in some tribal villages in our district till today. On the eve of the designated day, all the women and girls from the village, along with the animals (though there are exceptions), cross the River Surya and stay for twenty-four hours in the open on the opposite bank of the river. Only the men are allowed to stay back. The doors of all the houses are kept open. All night long the men sit under the huge banyan tree around the statue of Gaon Dev and intermittently beat drums and let out blood-curdling screams. This is to invite the gods to visit each home and rid them of any evil forces. The lamp near the god is to be kept lit and the men ensure that it does not go out all night.

An early morning walk through the village and I cannot spot a single soul except old Pavan Kaka hobbling to the riverbank to check if his wife has survived the cold winter night on the opposite bank. She was not keeping too well and had slight fever when she left for the opposite bank the earlier evening. The silence is broken only by bursts of crackers and the screams of men huddled under the banyan tree.

Over the years the patrolling by the village head has been done away with but the villagers still recollect those strict days. Sakharam recounts a time when his family had forgotten to take a rooster and for some reason it never crowed. As luck would have it, it crowed just as Rama Thakur was patrolling near their house. The bird was immediately killed with his catapult and his family paid a fine of four annas (twenty-five paise). The bird was a handy sacrifice.

Though rules have been relaxed and no one takes animals with them across the river now, the tradition of paying fines for violation still exists. One resident of the village had to marry off his son a week before the Gaon Dev ritual was performed. It was a love marriage and the bride's father could not listen to reason and refused to delay the marriage by even a day. The village council decided to levy a hefty fine of Rs 10,000 for violation. It was after much pleading and requests that they finally settled for Rs 4000.

The next day, as a sign of gratitude to the gods, animals are sacrificed. Last year there was enough money to buy five goats, one pig and a few chickens. The meat which is roasted on a wood fire is then distributed equally to all the villagers.

The elders of the village build a makeshift gate at the entrance to the village using two silk cotton tree branches to hold up a garland of mango leaves and flowers. The women who have spent a freezing night in the open are welcomed back home in the evening and enter the village through the gate. The dinner that night will be meat and rice.

While the entire village followed the rules set centuries ago, not everyone seemed to agree to it. Some of the villagers I spoke to felt that it was wrong to expect women to spend an entire night out in the open. Some even said that it was not required in today's world as they did not believe in ghosts and spirits. But the question none was willing to answer was, 'Who would be the first to oppose this and invoke the wrath of the Gaon Dev?'

Spells and Curses

In the year 2010, the villagers decided that they would not opt for the canal water. The land had been abused for a few years and they all decided that taking a break would be good for the soil. The canal authorities were informed and the water was stopped just before it reached the village.

While the refusal to use canal water may have been good for the land the villagers were faced with a serious problem. The groundwater level in the village dropped and soon the village wells started drying up. Women had to get up early in the morning and just managed to draw a vessel or two of water for drinking.

At the farm, it was groundnut harvest season and work was in full swing. One afternoon a woman sauntered in with a huge bucket full of soiled clothes and an empty vessel. I did not recognize her and asked her what she wanted. She rudely replied that she had come to wash clothes and take water. I said, 'Why here? You can go to the river and wash clothes.'

She turned around to leave. As she neared the gate she said to me, 'If you don't want to give water just say so.' I could not believe what she had said. The groundwater levels were dropping and this woman wanted to wash clothes using this precious water when there was a river flowing by.

Later I learnt that she was B's wife and was slightly eccentric. I dismissed the matter as an unnecessary irritant in the days of heavy harvest work. A few weeks after this episode which I soon forgot, I was faced with a strange problem. I would feel hungry and take food on my plate, but halfway through the meal I would start retching and could not eat. On most days half the meal was thrown away.

At first I ignored it, thinking it was the food or that I was not hungry and so on. But the problem just did not subside and soon Meena too noticed it. She remarked that I was wasting a lot of food which I rarely did. I was also losing a lot of weight and looked drawn and tired.

I got myself checked by the doctor and did the mandatory blood tests. All parameters were normal and there did not seem to be anything wrong with my body. But the retching and the reduction in food intake still continued.

I discussed this with Baban and his wife, Babita, who immediately said it was a curse and I would have to break it. A non-believer in voodoo or witchcraft, I dismissed her. As weeks passed, I was at my wits' end and finally agreed to try out the witchcraft option. Vaman Kaka from the village was an expert in breaking spells and his service was solicited.

On the designated day I reached Baban's house in the evening. Vaman Kaka made me sit cross-legged on the floor. He then proceeded with his voodoo spell-breaking ritual. He took a handful of rice and moved his hand all over my head and body. He then splattered the rice on a plate and stared at it for a few minutes, all the while chanting something. Then he asked me to close my eyes while he gathered the rice into a leaf. A few minutes later he wiped my eyes with water and sat in front of me. After ten minutes or so he broke his silence and told me, 'It's a woman's curse, but I have broken it. You will be fine.' I gave him Rs 50 as fees for his services and fell at his feet to seek his blessings.

In the meanwhile I learnt that C, B's only daughter, was getting married. C was a sweet girl who used to come to the farm to work and I was fond of her. The next weekend when we went to Mumbai I asked Meena to buy a nice sari and blouse for C.

On my return, I visited B's house and gave the sari and blouse to C. I explained that I was giving it before the wedding so she had time to stitch the blouse and wear it for one of the functions. I also included a small cash gift for her. As I was leaving the house B's wife fell at my feet and muttered, '*Maff kara, chuk jhala tar* (Forgive us if we have sinned).'

I thought it was strange for her to have done that, but dismissed the whole thing. A couple of weeks later I noticed that I had started eating well and the retching had stopped. A month later I gained 1 kilo. I could not figure out how the problem had vanished.

Many months later, as I mulled over the incident, I started connecting the dots.

Could it be that B's wife had cast a spell on me for my refusal to give her water?

Could it be that Vaman Kaka's spell-breaker had worked?

Could it be that C's mother had decided to break the spell as I gave gifts to C?

Why did she fall at my feet and seek my forgiveness?

I had no answers. All I knew was that my problem had been solved. I had no idea how it had happened. I resolved that whether it was true or not, I would be careful in my future interactions with the village folk.

Allahrakha, the baby barn owl

A butterfly on our bougainvillea plant

Chief and the Hen Log

Field beans for the day's meal

Harvesting pigeon peas

One of nature's colourful beauties

Plucking betel leaves for a village ceremony

Prize catch—a blue dragonfly

Rare visitor—the blue oakleaf

The shining krait

With Baban after a good harvest

Turmeric plant

Women harvesting the bumper moong crop

Working on *Moong over Microchips*

The Mahalaxmi Temple Fair

Every year, on the full moon day after Holi, the festival of colours, the Mahalaxmi Jatra begins at the base of the mountain with the cave temple. It lasts for fifteen days and is an event the entire village looks forward to. The crowds at the *jatra* are huge and on certain days run into lakhs of people. Busloads of people come for the jatra from as far as Gujarat. The deity is considered to be very powerful. The temple is only a few kilometres from our village.

As one enters the road leading to the temple, you can see stalls on both sides of the lane. There are a number of shops selling crockery, bangles, jewellery, clothes, vessels and even tin drums which are a hot-selling item. Meena too got one for herself. A little away is the kids section with a giant Ferris wheel, a merry-go-round and the circle of death where a man tries to defy gravity as he goes round and round on his motorbike in a globe.

The entrance to the temple is covered with rows of stalls selling sarees and coconuts and flowers and on certain auspicious days, one has to brave long queues. The temple itself is a small structure and the goddess stares at you with her big dark eyes set in an orange face.

The main attractions at the fair are the goods that can be bartered and the skewered meat fresh off the chopping block. Here again there are many lanes. The first lane is the dried fish lane with rows of stalls selling varieties of fish. Besides the famous *bombil* (Bombay duck), there are dozens of other fish that have been salted and dried. We cannot even recognize some of them.

The next lane is the blacksmith lane where one can buy all sorts of farming equipment like sickles, axes and choppers at a good price. Of course one has to haggle a lot to bring the price down.

The third lane is the onion and garlic lane. Here you get them at good wholesale rates. The highlight of this lane is the barter system. You can take any seed you have and it can be bartered for a range of items. You can barter seeds for onion, garlic, turmeric, sugar, dates and even salt. The forest department decides the barter rates and puts a board at the start of the jatra. An example would be that for every kilo of cashew seeds you got 12 kilo of onion or 1.5 kilo of garlic.

I spoke to the official at the government stall and asked him what they did with all the seeds. He said that the cashew seeds were processed into cashew nuts, while the other seeds were taken by the forest department for their various schemes. He also mentioned that some of the seeds have medicinal value and they are sold accordingly. I have a few cashew trees at the farm and each year we barter the cashews for various items. The onions I get from the jatra last almost till Diwali for us.

At the end of the dried fish lane is the highlight of the jatra, the *bhujing* section. This is the section where you get fresh mutton or chicken kebabs. Huge stalls are set up with freshly cut goats hanging from hooks. You buy the mutton and it is pounded in front of you on a tree trunk. Attached to the shop is the stall of the kebab maker who barbecues the kebabs on a charcoal fire and serves it. With so many charcoal fires the whole area is smoky and hot.

Our favourite stall used to be Zulekha Bibi's stall. The stall was the cleanest that we could spot in the entire lane. She lived in Dapchari, a few kilometres away, and was there at the jatra every year with her entire family. Her sons set up the mutton shop and pounded the meat for you. Her daughters did the task of making the kebabs and putting them on the skewers, while her husband sat at the fire and roasted the kebabs. Zulekha sat in the centre of the stall on a plastic chair and oversaw the entire operation.

Once the meat was pounded by the boys it was handed over to Zulekha on a huge platter. One of the daughters held up the tray of masala and spices to her. She took pinches of each spice and masala and added it to the meat. There was no measuring spoon or cup. Once she added the spices, she mixed the whole thing till it reached a consistency that she was happy with. The platter then returned to one of the daughters who made the kebabs before their father started slow-roasting them on the fire.

Meena and Zulekha got along well. She would invite Meena to sit next to her while the kebabs were being made and chat with her. She had already grilled Meena for every bit of information on us. In 2015, when she was told that Meena had quit her job, she said, 'Don't worry, I will find you a nice government job. I have lots of contacts.'

After the kebabs were ready, you could sit on the chairs at the back of the stall and eat. There were *pav wallas* too on cycles selling hot fresh-off-the-oven pav. We found the place too hot and smoky and usually packed our kebabs and rushed back to the farm for a sumptuous meal.

In 2017, we went to the jatra and walked up and down the lane looking for Zulekha. We could not find her or her sons. As we walked up and down one stall owner called and asked us what the matter was. I asked, 'Do you know where is Zulekha Bibi's stall?' He replied, 'I am her nephew. I am sorry but she just passed away last week.'

The news of her death was too shocking for us. We could not bear to eat kebabs that year and left the place.

10

Of Kerosene, Groundnuts and Subsidies

While it was a charmed life in some ways, apart from the hardships, we also had to deal with the corruption. Corruption was a way of life in the village. Be it the local sarpanch, the talati or the ration shop owner, each one was out to fool the villagers and extract money out of them.

My first encounter with this was at the local ration shop. Since I did not have a card in the village I asked Baban if I could buy some kerosene using his card. I needed it to light the lamps in the evening when there was no power. He gave his card saying that he was entitled to 10 litres but needed only five so I could buy the balance. At the ration shop in Dhamatne, the next village, the man said I could only get five as the stock was less. When he charged me Rs 60, I was puzzled since I knew that kerosene cost Rs 9 or 10 a litre. When I asked for the bill, he refused saying no one took a bill in these areas. I glanced at what he had written in the register and saw that it was less than Rs 10 per litre. It meant he had charged an extra

Rs 2 per litre. When I protested, he told me, 'It is none of your business. You got kerosene, that is good enough.'

I was enraged at this retort and told him that if he did not give me the bill I would complain to the officer. His reaction was a short guffaw, followed by a statement that he had been authorized by the officer to get more money for each litre. So the proceeds were being shared right up to the top. I returned to the village with kerosene that cost me Rs 60 while the official cost was Rs 48 odd. I called a few elders and explained to them how the ration shop owner was taking them for a ride and charging extra from each villager. They nodded wisely but told me they could not do anything about this.

The ration shop owner had a palatial house which he got painted every year. He had two sons who drove around in their jeeps and mostly behaved as if the village and its surrounding areas belonged to them. It was obvious that he had made a lot of money selling ration to the villagers since for every cricket tournament he would donate 50 kilograms of rice to be used for the food made for the players. It was as if he was atoning for the sins he had committed all year round.

The next incident was at the panchayat office where the entire village including me had gone to buy groundnut seeds from the government. The government was giving us a 50 per cent subsidy on each sack. It was my first time at the panchayat office and I just followed the villagers. We all stood in a long queue waiting for our turn to pay the money and collect the receipt. The receipt had to be shown at the godown next door where we could collect our groundnut seed sacks. When my turn came, a rude officer pushed a register towards me and asked for my thumb impression. I said I wished to sign. He looked up in astonishment and thrust a pen towards my nose. I signed and paid Rs 600 for a 30 kilo sack. I bought two such sacks. He did not give me a receipt but just a piece of paper on which was scribbled 'two sacks'. I noticed that the register where I had signed had nothing written on it except my name and the figure two.

We all went to the godown and collected our sacks. I was seeing them for the first time and examined them closely. I noticed that there was a small tag attached to it signed by an officer of the government of Maharashtra, which listed the name, date of manufacture, quantity and price of each sack. The price was listed as Rs 990 per sack. Considering that the government offered 50 per cent subsidy, the price we should have paid was Rs 495 per sack, but we were charged Rs 600. I went back to the officer who had taken the money and asked him about the higher rate.

Mr Y, the officer, looked up from his register and said, 'Can't you see so many people are taking it without questioning? If you don't want it, we can refund the money to you.' I just walked out of his cabin and went to the Block Development Officer (BDO) who was the final authority in these matters. At first, the sepoy outside tried to persuade me to not meet him, saying sahib was busy and suggested I return the next day. I was about to leave when I noticed a board with the timings when the BDO would meet the public and it happened to be the very day I was there. The timings also matched and there was no way he could refuse to meet me. I pointed out the notice to the sepoy and told him that if the BDO was busy in a meeting, then I was willing to wait till it got over. I had learnt a bitter lesson earlier at the pranth office and knew the ways and means of the people working in these offices. Had I left and returned the next day he would have shown me the very same notice.

The sepoy reluctantly went in with my card and within a few minutes, ushered me into the cabin. A young BDO was staring at a presentation on his computer screen. I narrated the entire story to him and asked for his intervention in finding out the right price for the groundnuts. He claimed complete ignorance of the whole episode and summoned Mr Y. He arrived within a few minutes and glared at me for a good minute or two before asking his boss what the matter was. The BDO questioned him about the price. He asked Mr Y how he had arrived at the figure of Rs 600 per sack. His explanation was stranger than the one Mr Z at the revenue department had offered his boss.

Mr Y claimed he had never seen a sack of groundnut seeds before and was not aware that the price was written on it. He claimed he had received a circular from Thane, which was the district headquarters, stating the price as Rs 1200 and so he was charging Rs 600. I quickly went out of the cabin and summoned one of our village boys to bring a sack of groundnuts into the cabin. I told the BDO that I wanted Mr Y to now examine the sack in detail. Both the BDO and Mr Y gathered around the sack and examined it as if they

Of Kerosene, Groundnuts and Subsidies

were seeing it for the first time. They declared I was right and the price was indeed Rs 990.

The next request from the BDO was to see the circular which had come from the head office stating the price as Rs 1200. Mr Y calmly told him that the file in which he had kept the circular had been sent to the collector for some other work and he did not have it with him right now. He left saying that he had a lot of work and there was a long line of villagers waiting outside his cabin for the groundnuts.

An embarrassed BDO offered me tea and tried to cover up for his subordinate's rudeness. He went on to explain how they were doing exemplary work in the taluk and invited me to see his presentation on the computer. I gently turned the topic back to the question of the price when he grandly declared that he would take it up with the head office since obviously the goof-up had happened there. He promised to keep me posted and in case there was a refund on the seeds, he would arrange to have it sent to the village directly. I invited him to our village and left.

Outside Mr Y had called a few guys he knew from the village and asked who I was and where I came from. He told them that I was raking up unnecessary issues and if this went on, the government would have to rethink the subsidy policy and may be forced to withdraw subsidy on groundnut seeds. A frantic group was waiting outside to ask me to slow down and let it be. After all it was only Rs 100 extra per sack that he was charging. I tried to explain to them that Mr Y had no control over the subsidy and these matters were decided at the highest levels. It was of no use since the entire village looked at Mr Y as the messiah who gave subsidies to them.

I realized the futility of fighting the system alone and returned to the village. The system had ingrained corruption into the villagers so much that they felt it was okay if the government officials charged a bit extra. After all, they were only taking a bit, not all of it.

The next day, while chatting with Moru Dada, I casually asked him how much he had paid for the groundnut seeds. It transpired that they had been charged Rs 630 per sack the very next day after we had picked up our quota. I wondered what could have happened in one night to increase the rate by another Rs 30. Probably, another circular from the head office that could not be traced!

A few months later I happened to visit the panchayat office to check if they had any other items on subsidy which I could buy. I met a most cordial Mr Y who saw me and said, 'Where did you disappear? You should visit us more often.' He then went on to offer me various items on subsidy like the sickle, the sprayer and many varieties of seeds. He even offered to enrol me in certain schemes where I could allegedly make money. He was kind enough to recommend a vendor close by who would provide me with a bill for a nominal fee of Rs 100. I just smiled at each suggestion of his and left the office in disgust. To ensure that I kept quiet he was trying to drag me into dishonest schemes in the guise of subsidy. Till date I have not received any reply from the BDO on the

Of Kerosene, Groundnuts and Subsidies

available subsidies and he makes it a point to avoid me each time he comes to the village.

The modus operandi of the government officers was very simple. Target the most influential person in the village and offer him some sop or scheme. Once he was bought over, he could not object when they went about looting villagers who were either too scared to raise questions or just did not have the time or money to follow up on these matters. A single visit to Dahanu would set back a villager by Rs 40 in 2004. Who would want to travel to Dahanu every once in a while to chase these officers? Everyone in the government knew this and used the simple tactic of procrastination to deter any villager who showed even a semblance of revolt.

Mr Y has been transferred and replaced by Mr K. When I went to buy the grass cutter in 2015, as usual, I was not given a receipt for the money I paid. My request for the same was met with a stony silence. Since the amount was big, Rs 15,000, I suggested to Mr K that I will pay by cheque or Demand Draft (DD). He replied, 'We do not accept cheques and the DD has to be from a particular bank. Besides you can collect the grass cutter only a week after you give the DD.' I tried to convince Mr K that a DD is as good as cash and there is no need to wait for a week. He said, 'It may be but we wait for a week.' I finally paid in cash. In this age where the government is trying to move towards a cashless economy, here was one department that insisted on cash payment for all purchases.

Moong over Microchips

My village also got its share of the famous Employment Guarantee Scheme (EGS) launched by the Maharashtra government. Under this, one person from the village was awarded funds for building roads or farm ponds. The government paid the money and the works were carried out by employing local people who earned daily wages as per the government norms. It was an excellent scheme on paper since most of the villagers had spare time after sowing and before their harvest and were on the lookout for work. One of the villagers built a pond in our village where fish were to be bred. The total amount given for the scheme was around Rs 40,000. He used machines to dig the pond and fudged the documents to show that people had worked for it. The question is, who benefited from the scheme?

Our village has eighty houses but the road to the village is still untarred and has gravel, which gets washed away every monsoon. The women in the village often complain that the stones cut their feet when they go to get water from the community well. In the next village there are only two houses but they have a tarred road right up to their doorsteps. I was confused at this anomaly and once asked the group sarpanch when I met him. He explained that these had to be taken up by a local person and usually the village has to contribute 10 per cent of the cost upfront while the government puts in the rest. Every road that is built results in a profit of nearly Rs 1 lakh as kickbacks to the contractor. He proposed that I take this up and put the initial amount and he would organize the rest. I shuddered at the thought of perpetrating another scam and decided that I would just buy better soled shoes to come to the village.

It was a shock that even basic amenities were not available to the villagers. Sitting in the city we read about the grandiose plans of the government to provide electricity and proper roads to every village by the end of the year. It was only when we reached the village and actually saw what was happening that we realized that these were just plans and none of it ever materialized. I thought of all the complaints that one had heard in the city of slow Internet connections and bad roads when here in this small village there was no road and telephone at all.

Anyway, two years after I bought the farm, it was a private telephone operator who bailed us out. Along with a few other villagers, we too got a phone, a wireless landline, though it worked erratically! A few years later the mobile revolution reached the village and soon there were multiple private operators who erected towers around the area. We now actually have a decent network and on some lucky days even 3G for data.

Scams Here Too

In the village there is no money in surplus. But for scamsters even schemes involving tiny sums are a good source to tap. One such was the Maharashtra State Electricity Board (MSEB) scam.

The MSEB distributed monthly electricity bills and they reached the villagers usually a few days before the due date. The nearest centre which collected money on behalf of MSEB is the Thane Co-op Bank in Kasa. The other option is to travel to Dahanu to the MSEB office and pay. Dahanu being far away, most of the villagers opted for the bank. As the due date approached, one can see serpentine queues outside the bank. Sometimes people spent the entire day there.

In 2016, a new centre opened in Kasa. There were hoardings everywhere saying that one could pay their power bills at this new centre. I went to check out the place. It was a neatly done-up shop where I met Amit Jadhav from Nanivali. He had installed two computers with a printer in between. Two young girls punched away in the computers while Amit collected the money. There was no crowd and the work was getting done very fast. I inquired about how he managed to start the centre. It was a franchisee for some businessman from Malad, Mumbai, who had received the licence to collect money on behalf of MSEB. He paid 1 per cent of the collection money to the franchisee. The firm sent their own people to pick up the collected money, besides providing the computers and printer.

It was not long before the centre became popular and the queues outside the bank started dwindling. Amit was happy for his collections were crossing approximately Rs 20 lakh each month which meant he earned Rs 20,000 a month. There was a lot of excitement and soon we heard of new centres opening in and around the village. Nanivali had its own centre opened by Amit and all the way to Boisar one saw boards announcing the bill payment centres.

Everything seemed to go well till one month some people found that the current bill did not reflect their last month's payment. They went to the centre and questioned Amit. He promised to check and get back and dismissed it as an error. Soon enough the number of people who turned up complaining that the last month's payment was not reflecting in the bills had shot up.

It did not take long to figure out what had happened. After the initial months of remitting the amount to MSEB the businessman had stopped paying. Amit, along with a few friends, went to Malad, Mumbai, to check the address they had been given. They just saw a locked office. When they checked with the neighbours they were told that the office had shut down a couple of months ago and they had no idea where the owners were. While they were standing outside the office they met a lot of other youngsters who had also come there to check. They were all in the same situation as Amit. It transpired that the scam was not limited to Kasa but was spread across the entire area from Virar to Dahanu. There were more than twenty-odd franchisees that had been opened to cheat people.

They all went to the police and lodged complaints. Months passed by but the police were unable to trace the businessman or his associates. The total amount of money he had duped from the unsuspecting franchisees ran into crores.

The centre at Kasa owed around Rs 30 lakh to MSEB which Amit had collected but had not been paid to MSEB. When they approached MSEB they were told that they had no idea about the franchisees that had been opened and they had only given permission to the company to collect money on their behalf. They had no idea where the owner was and what had happened to the company.

Amit had no choice but to refund some of the money to the consumers who were thronging his home in Nanivali. The case has still not been solved and the businessman is absconding.

Twinkle, Twinkle

Twinkle was a chit fund company extremely popular in our village. Jayesh the postman was the official agent for the company and did his work diligently. Each month you paid money to the chit fund with assurances that after six years you got double of what you had paid. The monthly deposit can be as low as Rs 500. You also had the option of parking large sums which will be doubled after the mandatory six years.

Jayesh had approached me too, asking for some money to be put into the fund. He showed me papers and brochures that spoke at length of the benefits of it. I did my checks on the Internet and found nothing that said the company was reliable. Some good sense prevailed and I did not part with any of our money. I also warned Baban and family that this did not look good and they should be wary of it. I suggested that they open fixed deposits in the bank which is more reliable and gave pretty good returns. In spite of my warning, Baban's son started putting Rs 2000 each month into the fund.

For some reason the villagers do not like going to the bank. They are scared of the paperwork and the rude attitude of the staff. Here Jayesh the agent came home each month and collected the money. They did not even have to fill any form; he did everything for you. The entire transaction was in cash which suited the villagers. Also the idea of getting double the money was more alluring than complex interest rates which they could not understand.

After the demonetization was announced, Vipul called me and asked about Twinkle. I asked him why he had asked. He said, 'A large number of these chit funds have collapsed as they dealt in cash only.' Sure enough, the next month, Jayesh did not come to collect the monthly deposit. When everyone asked him he just said there were some issues with the company but all will be resolved.

The company had collapsed. All those who had put their money in it now started hounding Jayesh asking for explanations. The situation turned so bad that he had to move to his wife's village for a few days till the heat died down. A few months later he went to each family and assured them that they would all get what they had invested without any interest. For this they would have to fill up a form. He promised that the money would be paid in a few months.

Forms have been filled and submitted but no one has yet got their money back.

Sarkari Troubles

While farming was becoming increasingly absorbing, the one source of trouble I had was with the government or sarkar, as people referred to it in the village. Right from the day I wished to buy land I had been at loggerheads with the administration. It seemed that corruption was rampant around us. Since we had already decided that we would not pay money to get work done there was no other option but to take them head on for each and every activity that we wanted done through them.

For a simple task like getting a copy of the 7/12 extract from the revenue department, I had to meet the official concerned and fight to get the work done. The villagers always commented that I was just wasting my time and energy since the same task could be done in a jiffy if I were willing to agree to their demands. My efforts to explain to them that these were due to them without any fees and the talati's existence as a revenue official was to serve the people did not find any favour with them.

After five years of farming, I was surprised when one day I was paid a visit by an officer from the Surya canal project. This project was completed around twenty years ago with the sole purpose of supplying irrigation water to the Adivasi villages in Thane district for agricultural purposes. Most of

Of Kerosene, Groundnuts and Subsidies

the villages around benefited from this canal and were able to grow two crops instead of one. Besides giving water to the villages the water was also being diverted to the industrial estate near Boisar. It was rumoured that the industrial estate paid a huge sum of money to the government for the use of this water.

Our land was on high ground compared to the surrounding land and hence there was no access to canal water. I invited the officer and offered him water and tea. After the usual small talk he told me the purpose of his visit. He claimed that since I was drawing water from the river I would have to pay the government for the use of water. I was shocked and explained to him that the river was a natural resource. The pipes were mine. The pump to draw water was mine. I had a meter and paid for the power to run the pump.

Since everything that irrigated the land was personal and the water was natural, why would I be required to pay the Government of Maharashtra for the water? What was their contribution to the irrigation of my land? He mumbled about some rule that had been passed by the irrigation department and he had no access to the same. I told him that I would find out and asked him what would be the kind of money that they would charge for a year.

His reply made it clear why he had taken the trouble of visiting me at the farm. He said it would be in thousands but was willing to reduce it if I 'cooperated' with him. This made me lose my cool. I told him to leave immediately and to serve me a bill of the huge thousands he was talking about. I also told him I would take the matter up with the authorities concerned and as of now would not part with a single penny.

After this episode there was no news from the Surya canal project office till almost a year later when a gang of five people landed up at the farm with a senior officer in tow. The officer was also from the Surya project and he explained that the money I would pay was for the government to recover the cost of the dam. I told him that this was acceptable to me but they would have to tell me how much and I would need a bill for the same. I was not ready to part with any amount without any paper evidence. He assured me that he would send a bill immediately and wanted to know how much land I owned so he could calculate the amount. I told him that the land I owned could not be the basis for the calculation since I did not use water on all the land. The usage was dependant on the crop and the area that was used for agriculture only. I had so many trees on the land which I did not water at all. Why would I pay for use of water on that land? I also explained to him that the consumption of water would vary according to the crop I sowed and they cannot charge me a fixed rate based on the area that was irrigated.

The poor man had no answer to my questions. He said they did not have the mechanism for this kind of detailed analysis but he agreed to measure only the part that I had sowed and calculate the amount based on that. His team went around the farm measuring the land that had been used. He promised to send me an official bill which he said would arrive in fifteen days.

Of Kerosene, Groundnuts and Subsidies

Almost a year later, one of the villagers, Santosh, called me and handed over a bill for Rs 1700 from the Surya project office as dues for water usage. They had calculated the use of water from 2000, when I had not even bought the land and arrived at the figure. I showed the bill to Meena and asked her to inquire with the irrigation department. The bill had no details nor workings on how they had arrived at the figure. I was ready to pay the money if they agreed to share the calculation with me.

She went to the irrigation department in Mumbai. They all laughed at her and told her that in their entire career in the department this was the first time that a farmer had come asking for calculation details and wanting to pay. They advised her to keep the bill safely and forget it. We had no idea what the scam was but we decided to pay the bill. Since then no official has visited us and no other bill has been served.

In 2008, there was a cyclone which wrecked crops in the state and the government decided to compensate the farmers for the loss. The procedure was that the talati and the agricultural officer would visit the village to assess the damage and then pay the compensation. I kept asking Baban if any officer had come but there were no visits. A few months later Baban told me he had got compensation from the government. I went to the sarpanch and asked him why my name was not on the list. He promised to find out.

A few days later, he explained that there had been some mix-up and my name did not figure in the list eligible for compensation. I was furious at being left out as I was a legitimate farmer and did lose crops. He calmed me down and said that he would ensure that the next time I would be included and to forget this year's compensation. I agreed to not raise the issue and gave him a copy of my 7/12 extract so I could be included in the next round, if any.

Two years later, in 2010, there was excessive rain and most of the rice crop, including mine, got washed away. We barely got enough rice to store as seeds for the next year. The government again rushed to the rescue of the farmers and announced a compensation package. This time I went to the sarpanch and reminded him that I had to be included. I had lost all my rice and I too was a farmer in the village like them. He agreed to inform the authorities and include my name.

A year later the entire village received the compensation except me. On inquiring, the sarpanch expressed his inability to do anything but claimed he had given my papers to the field officer of the agricultural department. This time around I was not going to take this exclusion and needed answers.

I went to the agriculture department near Dahanu and met an officer there. I told him my problem and also showed him a copy of my 7/12 extract. He flipped though some register and said he could not find my name. I asked him who would have the answer on why I was excluded. He calmly told me to contact the talati since he was the one who had made the list. I immediately went to meet the talati at Kasa.

He was not around but the circle inspector, his senior, was willing to help me. He explained that his job was only to provide the list of land records and the agriculture department was the one which prepared the list eligible for compensation. I requested him to show me the register so that I could confirm that my name did indeed figure in the list of people having land in Peth village. He showed me the register with my name but refused to allow me to take a photocopy of the same. I realized that this was another game of passing the buck being played by both the revenue and the agriculture department.

I decided to take the matter to another level. I wrote a long letter to the agriculture minister and the agriculture secretary at Mantralaya, the state secretariat, and went and gave the letters personally to the departments. I got signed receipts of the letter. The letter asked for an explanation of why my name was not included in the list of farmers eligible for compensation due to excessive monsoon in 2010.

I waited for a month to pass, knowing well that there would be no response from the two offices. A month later I filed a Right to Information (RTI) plea with the department concerned, this time asking for details on the action taken by them on my letter and a request for information on any such action.

My RTI plea had the desired effect. I got mails from the department saying that they had forwarded the matter to the Thane district headquarters and I would hear from them soon. A few days later a letter arrived from the Thane district headquarters saying that they had sent it to Dahanu for further action. So they were keeping me updated on each action they were taking on my letter.

The very next week I got a call from Khare, the taluk agriculture officer. At first he was aggressive and told me that I should not have written to his seniors but approached him first and so on. I listened to him for some time and then calmly told him that I had filed an RTI and it was his job to reply to the same, not call me. I also mentioned to him that time was running out since as per the RTI Act he had to reply in thirty days or face penal action for not replying.

The next week I was at the farm and Khare, with a large team of six people, came to meet me. These included the agriculture officer of the village, the sarpanch of our village, the taluk officer and the nodal officer. They sat down and the agriculture officer of the village, Gharat, made the introductions. At the end of it I asked him who he was. He said he was our village officer. I turned to his superiors and smiled and told them that for the last eight years I had been here the local village officer had had no time to visit my land. I did not even recognize him. How on earth would he know what was the damage on my land and how much compensation I would receive?

The tone for the interaction had been set. They spoke at length on various schemes and benefits that they wished to bring to the village and how they would transform the entire agriculture. I listened patiently and then told them that I had never heard of any scheme of assistance from the department in the last eight years. I could see Gharat fumbling and turning red in the background. Besides I told them that the purpose of the visit was not schemes but my

Of Kerosene, Groundnuts and Subsidies

exclusion from the compensation list and it might be better if we discussed that.

The officers tried to impress on me that it was incorrect to file an RTI for such a trivial matter and it would have been better if I had met them directly. I was waiting for this suggestion, and immediately told them how I had been to their office and had been misguided to the talati's office by none other than the agriculture officer.

I explained to the officers that this was not a question of money but a matter of service and incorrect data being generated. If the real assessment had been done and the visit to the fields taken place they would have included my name.

They all looked sheepish and knew that I was right. They had done no visits and had just picked up the previous year's list and sent it across. Finally they asked what solution I wished for. They hinted that they would include me under some scheme if I withdrew my RTI plea. I refused to do any such thing and told them that there was no question of accepting any other scheme in lieu of the 2010 compensation.

I told Khare that it was a mistake by their department and they would have to accept it. He asked me to reconsider my decision since it would mean a major embarrassment to his team. He hinted that this procedure would take a lot of time and it may take months for the money to come. I refused to back down and told him that I had waited for two years for the money to come and a few more months would hardly matter.

They left after walking around the farm, but not before one last-ditch attempt by the sarpanch to request me to withdraw the RTI plea and forget it. I asked him if he was willing to be left out of the compensation. I told him not to get involved and let me do what I thought was right. After they left, I wondered why the sarpanch was so keen on seeing me withdraw my RTI petition. Could it be that he was in cahoots with the agriculture department and had taken what was due to me?

It was a joy when after three weeks I got a call from the department saying that my compensation had been calculated and it would be transferred to my bank account soon. They wanted a letter from me saying that I was happy with the solution and my RTI plea was closed. I told the officer that when I saw the entry in my passbook I would surely send him a letter, but nothing till then.

The question was not about the money but the effort I had to put in to get what was due to me. There was no central method of information dissemination and everyone relied on the field officer to tell them what was due to them. I wondered how a poor farmer would get his or her dues if they had to go through so much red tape and paperwork.

Murder

Right from the start, I felt I would have a problem with Ramesh, who was an alcoholic. His house was on the side of the road as soon as you entered the village. Our paths would cross many times as I went to the farm or returned from there. His elder son, Jitesh, worked in Boisar and visited his parents rarely while the younger one, Bunty, was a sweet boy who had failed to clear his Secondary School Certificate (SSC) examinations and worked in a factory close by.

Whenever I met Ramesh, he would always greet me with his bloodshot eyes and ask me if there was any work at the farm. I politely refused his help. He would then launch into an explanation of how good a worker he was and how he needed money since his wife was ill and had to be taken care of. I was aware that all he needed was a quick buck to quench his thirst.

Ramesh drank all day and had raging rows with his wife whom he beat up very often. The village tried to reason with him but things did not change. One evening when the day was drawing to a close, he swaggered to the farm where we were all sitting and having tea. He demanded that I give him money. I could see Baban frantically gesticulating from behind telling me not to. I told him that he could work at the farm for money but I was unwilling to lend him any.

He suddenly became aggressive and said he had been trying to get work but I always avoided him. Now he wanted money which he said he would repay later in the form of labour. I refused to budge and politely told him that unless he worked he was not getting any money from me in advance. He raved and ranted for ten minutes before leaving. As soon as he had left everyone heaved a sigh of relief. He was not the first person to ask for money but his demeanour really put me off. Meena warned me against him saying that he looked like a violent man and must not be encouraged.

I liked Bunty a lot and he was welcome at the farm. He was soft-spoken and never even mentioned his father and the trouble they had with him at home. I came to know of his frequent fights with his father when his mother came with his food one day to the farm. She explained that he had left early without eating a morsel after having a fight with his father. There were days when he would come and just loiter around the farm, chatting and helping me with some sundry tasks after school.

Immediately after failing to clear his class X exam, Bunty left for a job in a chair-making company close by. He now visited the farm only when he returned for a holiday or some festival. There was little I saw of his parents till that day when Ramesh came asking for money.

A couple of weeks after that, things reached a brutal climax in Ramesh's life. I reached the village to find a convoy of policemen. Slowly I learnt the gory details of the event. Jitesh had come home for a break to be greeted with the sight of blood all over the house. His mother was lying on the bed hacked to death. The poor boy almost fainted right there and barely managed to crawl out of the house for help. The police were now looking for Ramesh who was missing.

A few hours of searching and the missing blood-covered axe was discovered among the bushes near the river. A few feet away the police found Ramesh hanging from a tree. He had used a thin nylon rope to end his life after the gruesome crime he had committed. Everyone in the village was numbed by the incident and it took us days to overcome the shock.

This murder and suicide came as a jolt to all of us. Somehow I had naively imagined that such violence was never possible in the village. At any rate I had never expected such brutality. I had lived in a city once teeming with gangsters and shootouts. Dawood Ibrahim, Chota Rajan, Varadarajan and others were household names. We had serial killers like Raman Raghav, but they were just names we read in the paper. We were not close to them or the crimes they committed. Life was cheap in the city. But somehow I felt personally involved and shocked at Ramesh's death. Things were too close for comfort here.

The gruesome crime was a major topic of discussion. Some of the villagers even offered to sleep in my house at night in case I was scared that Ramesh's ghost would return to the farm to haunt me. I jokingly told them that I would offer it alcohol which will ensure that it did not harm me. In spite of the jokes and the brave front I put on, the first couple of nights after the incident were difficult to pass. I also started locking up the main gate at night.

Thefts

In a small village with the houses so close to each other one would think that thieves would not dare come. But the village had its fair share of thefts during the last few years.

Jayendra ran a mobile recharge shop and everyone knew that at the end of the day he returned home with a lot of cash. In the middle of one night he heard his mother, who was sleeping in the next room, scream in panic. He rushed out to see the fleeting image of a man opening the front door and running out. Torn between checking on his mother and running after the thief, Jayendra decided on the former.

The thief had removed the tiles on the roof and jumped down. His mother was in pain for he had fallen straight on her and when she screamed he had run out of the house. As usual they did not report the matter to the police. The logic was that since he had not got anything, why report!

Once Jayesh the postman's wife entered the kitchen early in the morning and got the shock of her life when she saw a man sleeping in a corner. She went back to the bedroom and woke up her husband. Jayesh, instead of calling for help, foolishly took a stick and prodded the man. The man got up with a start, pushed both of them aside and ran out of the front door.

A quick check of the house and they found that he had entered through the window of the grocery store that they ran on one side of the house. He had gone through the contents of the cash counter and cleaned it. They went to the kitchen and found that the thief had even eaten the leftovers from the previous night. What no one could figure out was why the thief chose to sleep in the kitchen knowing well that the owners were sleeping in the next room. Again the matter did not get reported.

A few weeks later we heard that similar incidents had happened in the next village too where the thief had entered the house, taken some money and also eaten the food. It looked like the work of the same person. Someone finally had the sense to report the thefts to the police. Soon enough we heard that the police had caught a drug addict who was roaming in the area. He needed money for his daily fix and was targeting houses at random. Everyone heaved a sigh of relief to know that he had finally been caught.

We were not comfortable with the thefts happening so close and felt that the time had come to strengthen our fence and try and make it difficult for people to enter. The barbed wire fence was rusting in places and it was easy for anyone to lift the wires and sneak in. We got the fence redone so that entry would be difficult.

11

Doctor in the House

While basic amenities were lacking, the most serious lacunae were in healthcare. I learnt this the hard way. For some time I had not seen M and heard that he was ill. I went to visit him and found that he was almost half his weight and looking very pale. Not having much faith in the self-diagnosis that he had jaundice, I told him that he should visit a doctor. He agreed and the next day I offered to drop him to the doctor.

We reached the clinic or what seemed to me like a small hut, near Charoti, where there was a dirty bed and a pot-bellied doctor. The doctor asked him a few questions and then went on to administer two injections, one in each buttock. I was stunned when I saw him pick up the syringe and plunge it into a vial without even sterilizing it. He charged Rs 50 and gave no prescription or medicines. His only advice to M was to eat chicken.

On the way back I explained to M the need to sterilize needles and how dangerous it was not to do so. M said that since he had used two separate syringes, it was fine and nothing would happen. Besides, he explained that since the injections were administered on separate sides of the buttocks the chances of infection were very low. I realized the futility of my explanations and gave up. I was having lunch when Lahu Kaka's son Harishchandra came running and told me that M was having fits and was not registering anything. I left my unfinished lunch and rushed to his place to find M in a delirious condition with his eyes half-closed. I said that he must be taken to the hospital at once. I was shocked when his family said I would have to wait till someone got some prasad from the next village, which would miraculously cure him. I lost my temper and told them that I was taking him to the doctor.

My defiant tone did the trick and soon a couple of guys with M's wife got into the car and we shot off to the clinic with a moaning M in the back seat. The doctor had left but I caught the doctor's assistant and demanded to know what M had been injected with in the morning. Apparently the doctor had one standard cure—Neurobion and vitamin injections—which he gave everyone who came there.

With no doctor around, our next choice was to take him to a hospital. I suggested the Kasa primary hospital which is just a couple of kilometres away. M's wife and her brother were clear that under no circumstances would they take him to the Kasa hospital. They explained to me that there were no doctors in the hospital and it was run by ward boys and nurses. We had no choice but to take him to Dahanu which is around 30 kilometres away. We managed to reach Dahanu and admit him to a private hospital. The doctor there turned out to be from Mumbai, where he had been practising till he shifted to Dahanu to set up the hospital. He put M on saline drip and immediately recommended some blood tests. The next day he informed me that M had typhoid and pneumonia. There was no sign of jaundice. Since he had suffered fits, the doctor also advised a CT scan at a later date. The nearest scanning centre was only at Vapi, around 50 kilometres away, or in Mumbai.

A few days later, M's wife and her brothers came to pay me for the petrol since I had driven so far with her husband. When I refused to accept the money, they were surprised and told me that they thought all city people were only after money and never did anything free. I felt embarrassed that this was the image we had built in the minds of the villagers.

It was much later that I learnt that the doctor we had been to in the morning was, in fact, a quack and did not have a real degree. Years ago someone had died under his treatment and his licence was revoked. He, of course, continued to practise in his little hut near Charoti. The hospital at Kasa had been built a couple of years ago with the best infrastructure and facilities. Unfortunately there were no doctors since no one wished to take a posting there, even though it was only 100 kilometres away from Mumbai. Even if there were doctors there were usually no medicines available to be administered to the patients. Such was the pathetic medical infrastructure near the village.

Besides the 'double-dose' Dr S there were a few other doctors in Kasa. Most of them had BAMS degrees, which meant they were Ayurvedic doctors, but they blatantly prescribed allopathic medicines and even administered injections. One of them had even set up a small nursing home with five beds and gave saline to whoever chose to come to his hospital.

It seemed strange that the village folk rushed to the doctors for even small ailments. I could remember during our childhood whenever we fell ill my mother would always try out some home-made cure before taking us to the doctor. In the village they just seemed to have forgotten these cures or did not trust them. One night, in 2015, B's mother rushed to the farm and said her son was terribly ill and had to be taken to the hospital. They did not wish to go to the government hospital and chose a private hospital near Dahanu. I spoke to the doctor after he had examined B and the doctor said it was a case of acute acidity and nothing else. I questioned B and gathered that he had not eaten the whole day and had drunk a couple of beers with friends in the afternoon. The doctor kept him for a day and sent him back with Gelusil and instructions to eat before drinking after charging Rs 3000. I thought it was a waste of money when this ailment could have been treated at home itself. But then, as a policy, I never gave medical advice to people and it was up to them to decide the best course of action for their illness.

The Death of Moru Dada

One of the victims of the pathetic healthcare in the village was none other than Moru Dada. After our initial fracas over the tractor and the distancing of Moru Dada from our daily farming activities, we became good friends. We never raised the topic of the tractor and it was as if it had never happened. Almost every alternate day I would run into him somewhere or the other and he always stopped and inquired after us.

A couple of years at the farm and he too, like the other villagers, realized that I was here to stay and not running back to the city. Moru Dada was very helpful as having been the sarpanch till 2010 he knew of every scheme that the government announced and was always willing to share the information with me. He even shared saplings which he would get from the forest department or the agriculture department and guide me on how to plant them.

A few years ago, somewhere around 2011, he called me one day and asked me to drop by when I had the time. He did not sound very well and that evening I went to his house. He looked extremely ill and tired. We sat on the bench in his front porch and after the initial pleasantries he asked, 'Do you know what diabetes is?' On asking who had got it he replied that it was him but he could not understand what the doctors were telling him.

I gave him a detailed explanation of the disease and also stressed on the fact that he would have to be careful all his life as it was not going to vanish. I told him, 'Moru Da, you have to walk every day.' He replied, 'Walk where? I don't go anywhere now.'

I explained the need for him to exercise and that walking was the best form and it would control his disease too. He seemed to get what I was saying and said he would start the next day.

A few days later, I met him on the road, not walking but on his bike. I said, 'Arre, you are to walk everywhere and here you are back on your bike.' He smiled at me and said, 'Come home, I will tell you what happened.'

It seemed he had got up early and tried to go for a walk. The first few minutes were fine till he met someone from the village who stopped him to ask where he was going. It was indeed a strange sight for most villagers to see Moru Dada walking and that too so early. He said, 'I can barely walk for five minutes before being stopped by someone and you tell me to walk for an hour?'

I realized that it was not going to be an easy task for Moru Dada to walk. He was the former sarpanch and a political leader and there was not a single person who did not know him. Besides, the concept of a walk was alien in the village. You walked when you had to go somewhere for work or to meet someone. Why would you walk just for the sake of walking? I could think of nothing to advise Moru Dada other than to walk a bit early to avoid people.

A few days later, as I was sipping tea at Konduram's tea shop, one of the old men sitting near me remarked, 'Do you know that Moru Dada has lost it? *Toh veda jhala* (He has gone mad).' I looked at him in shock and asked him how he knew. He said that Moru Dada walks at weird hours and does not go anywhere. He just roams around without a reason and returns home. It must be a mental illness. I tried to explain that maybe he was just going for a walk as advised by his doctors. The old man replied, 'We have such excellent medicines now. Which doctor would have told him to just walk?'

On my way back, I dropped to check on Moru Dada. He saw me and said, 'So you have come to meet the mad man.' We laughed at the whole thing and I left after telling him to ignore what people were saying and continue his walks. I knew he would not do it and this would end in a disaster. Sure enough, within a month his sugar levels shot up and he was admitted to the hospital. He returned after a few days and this time the doctors prescribed insulin injection twice a day. He had to be put on a strict diet of rotis and vegetables with meals every two hours. His wife did all that she could to make him stick to the diet and nurse him back to health. He seemed to get better after a couple of months till tragedy struck again.

It was his wife this time. She had been complaining of headaches for quite some time and finally it was diagnosed as brain cancer. Moru Dada and his son took her to Mumbai for treatment. Meena met them in Mumbai and contacted an oncologist she knew to find out what we could do for her. Unfortunately the cancer was at an advanced stage and all the doctors advised that she be taken back to the village and kept happy till her final time came.

Moru Dada couldn't believe it. His world was shattered. He had taken her to Mumbai in the hope that something would be set right but he brought her back home after a month with little hope. After they returned, he called the local witch doctor, Shankar, in a last attempt to save his wife. Shankar, after a lot of mumbo jumbo, declared that evil spirits resided in the two tamarind trees outside Moru's house and they would have to be chopped down.

Moru Dada immediately got the trees chopped down. He would do anything to save his wife. The chopping of the century-old trees did brighten up his home but did nothing for his wife. She died peacefully surrounded by her family and friends within a month. This was the end for him. He wept for days openly and looked like he would never get over it. From a huge bulky man he had become a thin shadow of his former self and it was difficult for some to even recognize him. In his time he was a political fixer, a man who could get things done, someone who was willing to push all boundaries if it suited him. From a farmer he had become a fixer, a man you went to for organizing votes, people, meetings, almost anything. There was nothing he couldn't do for money.

All that came back to haunt him as he sat in the porch, ruing his life. People in the village had never seen him walk; he was always on his motorbike, a sign of his power and wealth. A few years ago, he had helped the local MLA win the elections and the MLA had gifted him a second-hand Maruti 800. Moru Dada tried driving the car but he was not happy sitting inside a box. The bike was his chariot and he was rarely seen off it except at home. But his wife's death hit him hard. His sugar levels shot up and he went to the hospital again for a while.

He came back completely broken. He kept telling me, 'I grow such good rice and all they give me is dry rotis. Death is surely better than this existence.' He stopped walking and started riding his bike. One day, while riding back from Palghar, his slippers fell off but he did not notice it. He had kept his foot on the hot silencer and by the time he reached home his toes had been burnt. Due to diabetes his toes were numb and he did not even realize that something had happened. The wound did not heal due to diabetes and finally the doctors had to amputate the toes to make sure that the gangrene did not spread.

Moru Dada had three sons and he had made every effort to see that they were settled comfortably. The eldest, Sunil, was interested in tilling the land. Moru Dada built a house for him on the land itself and got a tiller and other equipment that he needed. The second son, Nilesh, wanted to work in a company. Moru Dada used his vast influence and got him a decent job. His wife is a trained nurse and using his fixing skills he managed to get her a job too. The last son, Sachin, is a driver. Moru Dada got him a truck through a government scheme and even managed to get him a regular contract to deliver the Public Distribution System (PDS) ration for the government. After his wedding, Sachin left the village to settle down in Boisar.

Now, with Moru Dada on the bed, we heard rumours that the sons were squabbling over the huge medical expenses they had to incur.

One day, after his return from the hospital, I got a call from Nilesh. He said, 'Baba is not eating or listening to anyone. Can you please come and talk to him?' Baban and I went to visit him. He had stopped eating completely and was lying on the cot all day and night. He could barely sit up and speak and kept dozing off in the middle of the conversation sometimes. It was a sad situation and we could do nothing about it. I asked him why he had stopped eating the medicines. He smiled and replied, 'I do not want to be a burden for anyone. Besides the medicines are not helping me.' Obviously the squabbling had not gone unnoticed by Moru Dada.

I offered to buy the medicines for him. He said, 'So you want me to be indebted to you in my next life too?' He then offered me a raw jackfruit from his tree and said, 'You Madrasis make a nice bhaji out of this. Go and give it to Meena.' We made the raw jackfruit bhaji and I took a small portion for Moru Dada. He sat on the bed and ate a bit of it. He said, 'Now that I have eaten Madrasi food there is nothing left in this life for me.' I laughed and replied, 'There are hundreds of dishes you have not tasted. Wait till I bring you all of them.'

A week later, I got a call in the morning from Baban. Moru Dada had committed suicide. We rushed to his house to find hundreds of people gathered there with women wailing inside the house. We paid our last respects to him. Raajen Singh was also there. He had worked with Moru Dada for the welfare of the tribals. Besides they also had their marriage solemnized on the same day in the village.

Raajen and his wife, Archana, were very close to Moru Dada and the family.

As we walked solemnly to the cremation area near the river we noticed that there was no police. We asked someone why the police had not been called in since it was a case of suicide. We were told it was too much of a headache to have the police called and besides they would insist on a post-mortem, which would delay the funeral. It was better done this way where the case was not reported at all. Anyway the whole village knew he was ill and dying and it made no sense in dragging the authorities into this.

It took us a while to get over his death. There are so many people who live with diabetes and manage to control it. My mother has been having insulin for over thirty years now and she manages her diabetes very well.

A couple of months after his death, Sunil met me and gave me a ripe jackfruit from their tree. He said, 'Baba wanted you to have one. Please take it.' I saved the seeds of the jackfruit and in the monsoon planted it at the back of the house. The tree is now almost ten feet tall and growing well. It reminds us all of Moru Da the leader and also the good father, husband and friend.

12

Snakes, Owls and Other Animals

The Hissing Cobra

Staying right at the edge of the village with a river on one side and surrounded by fields and forests was definitely romantic. The romantic setting was also the stage for some unexpected encounters.

The paddy fields around have a lot of snakes and I have spotted the buff striped keelback, the rat snake, the krait, the deadly Russell's viper and cobra too. The river and its small islands are full of water snakes. The evenings are a treat as far as moths and insects are concerned. Every inch of the wall, especially near the light source, is covered with varieties of moths. Some of them have the most intriguing designs and I especially try and photograph the bigger ones. I have also spotted a couple of rabbits and a mongoose on the farm during my routine walks. Every once in a while we have scorpions as house guests—they have to be evicted gently.

I once had the privilege of seeing a buff striped keelback eating a toad during the monsoon. I shot the entire sequence which is a treat to watch even after so many years. Geckos roam the walls and with such an array of tempting bait, they are kept busy. They come in different colours and patterns. The ones we see in the city are a far cry from the beautiful ones at the farm. It is a treat to sit quietly in the evening and watch the geckos prey on unsuspecting insects. Sometimes, I would witness a territorial dispute between two geckos, usually ending with one of them scampering away all bloodied and bruised. A monitor lizard was a resident of the pump room for many years. It was curled up on the wall and it took me a while to figure out what it was, never having seen one before.

These reptiles and insects did not harm us in any manner and we enjoyed their company. I even had a tree frog on the kitchen shelf for months. The frog had been around for so long that it was almost like a pet and we named it Buzo. It often surprised us by leaping out of coffee mugs. Now we have frogs under the sink, jumping around at night looking for food or on the porch nestling under our slippers or shoes. They are very fond of Pepper's water basin and treat it like a personal Jacuzzi—floating in it till we evict them. These and the giant wood spiders which spin huge webs across trees are like pets and we look forward to them every season. What we were not prepared for was our surprise visitor.

It was Independence Day, 15 August 2005. Meena took a couple of days' leave and came with me to stay at the farm. We reached the farm around afternoon and as usual I started the mandatory check in the house. I looked under the cot, the table and behind the drums. As I entered the kitchen I got this eerie feeling that something was amiss. From the corner of my eye I detected a slight movement on the kitchen shelf.

I promptly told Meena that there was something in the shelf and asked her not to go near it. As fate would have it, there was no power, so I shone the torch into the shelf. Our guest reared its hood and started hissing. Its hood and markings clearly proclaimed what it was—a beautiful glistening spectacled cobra. We were too shocked to react; we just stood there like statues, with our mouths open. The first thing I thought of was to click pictures. I crept back to the car to get my camera. Every time the flash popped, the snake darted towards us, hissing.

We had seen cobras on television but this was something that we just could not get over. The hissing itself was sending a chill through our bones and besides it kept darting towards us. Luckily it was on the shelf, high above the ground. A few pictures later the camera batteries gave up. Meena offered to go to the village and get some help. I was not keen as our neighbours were scared of snakes and would be of no help. I had seen their reactions a couple of times when snakes had been spotted. They usually ran a mile from it.

We could not think straight, and decided that a cup of tea would stir our brains into action. It was an unforgettable experience making tea with a cobra hissing in the background.

The shelf with the cups was next to the snake and I had to quickly open the cupboard and take out the cups. The hissing only got more furious. Under the watchful gaze of our guest we made tea and stood in the corner of kitchen sipping and thinking of the next course of action.

Meena came up with an idea. She suggested that we keep a bucket under the shelf and then push the snake into it. Then it would be an easy task to lift the bucket out of the house—a silly idea from the start. We cleared the vessels and other things lying on the floor of the kitchen. This was to ensure that our guest did not creep behind any of them. Then we got a bucket and pushed it under the shelf.

I prodded the snake with a long bamboo stick. It hissed more viciously and darted towards me. I lightly kept the stick on its head and suddenly the snake went quiet and put its head down. For a moment I was tempted to catch it but thought the better of it.

A gentle prod with the stick and I managed to push the snake off the shelf. As fate would have it, it fell outside the bucket. Now there was a slithering, hissing, darting cobra on the floor of the kitchen. I was standing there with a long bamboo trying to push it out of the kitchen door while Meena was outside hoping to shoo it away. The floor was too smooth for the snake and it kept slithering and moving in the wrong direction. All this while it was hissing madly.

Finally, after a few tries, I managed to push it out of the door. Our guest had no intention of leaving and kept trying to return to the kitchen. There was no option but to flick it

Snakes, Owls and Other Animals

out of the porch. I was reminded of the *gilli danda* game I had played as a child when I flicked the snake into the field. It had taken us all of forty-five minutes to evict our guest.

It was a major relief to see the snake slither away into the grass. We both collapsed on the steps and just sat without speaking a word for the next fifteen minutes. Both of us were thinking of the consequence if the snake had bitten one of us. What could we have done? The nearest hospital was at Kasa, 10 kilometres away, and we did not even know if they had the antivenom.

Later in the evening we sat on the porch and carefully went over the events that had happened a few hours ago. We realized that it was one thing to see an hour of snake show on television and another to be confronted by one at home. We made a check to figure out how it could have entered the house. The errant branch of the mango tree, which we assumed was the way it had got through, was chopped down. We put steel nets on the windows of the house.

It also dawned on us that it was not the snake that was the intruder but us. We had encroached into their jungle and set up a house. How would the snake know that this was not a place for it to be? We respected the snake and did not harm it. Maybe that was the reason it did not harm us too. When the village came to know of this incident, they insisted we break a coconut under the jamun tree outside the house. If breaking a coconut could prevent cobras from being house guests, I was all for it.

It was more than two years later that I got the same eerie feeling when I entered the house after a weekend in Mumbai. I knew something was amiss. I did the routine check behind the door and under the table. All seemed clear. Contrary to my usual practice of going to the bedroom and changing I changed into my shorts and T-shirt in the living room itself. A quick check of the kitchen revealed nothing, yet I could not help feeling that something was in the house.

I opened the back door to let some light in (no power, thanks to MSEB) and went into the bedroom to check. I peered behind the cupboard and looked under the bed only to find nothing. Just as I was about to turn around and go about my chores I sensed a slight movement under the cot. There was definitely something there. I bent down and managed to see a small shiny tail vanish under the swing which had been dismantled and kept under the cot. I went and got our *waki* or snake stick. A gentle prod to the swing and out popped a shiny, small black head. This was positively a snake.

The first thought that came to my mind was that it looked very familiar. It reminded me of Michael Phelps the swimmer, minus the ears. Setting aside my Olympic thoughts I focused on the problem at hand. We both stared at each other and before I could be hypnotized I gave a quick knock on the wood.

It started moving to the other end of the room. I kept staring at it, watching in horror as it slithered across the room. By the time the tail emerged the head had reached the other end of the bedroom. It must have been a good eight feet long. I concluded it was a rat snake (*dhaman* or *aandhla*) as it had a shiny, black coat, no hood and a tapering tail. Thankfully these are non-poisonous.

Slithering (had a quick christening) was now coiled at the far end of the bedroom watching every move I made. I mentally wished it a good day and requested it to move out since it did not belong here. To my surprise it started slithering towards the door. It reached the cupboard and without even a moment's hesitation turned towards the door. Seconds later it was under the fridge in the kitchen. I could not believe this. I had not even prodded it to do anything. Could it be that I had managed to communicate with it?

I meekly walked behind it to the kitchen where it was peering at me from the side of the fridge. I realized why it had stopped. There was hardly any space under the door for it to crawl out. I moved slowly to the door and shut it halfway. Thanks to the uneven floor now there was a gap wide enough for Slithering to move out. I almost imagined a slight nod from Slithering as it went out of the door into the field. I hissed a gentle goodbye to it before shutting the door and stuffing cloth into the gap in the door.

That night I could hardly sleep; I was too excited. Could it be that I had finally managed to communicate with snakes? Why did it go out so meekly? How did it know that it had to go out of the door? It was too complicated to figure out. My thoughts were shattered by a loud thud. I got up thinking, 'Now what?'

I stood in the darkness in the kitchen trying to figure out where the noise was coming from. I moved gingerly to the bucket near the shelf which seemed to be the origin of the noise. I shone the torch into the bucket to find a small shrew (*chuchundri*) trying to jump out. Ah! So this was Slithering's dinner. I concluded that it had probably come inside chasing the shrew. Anyway, I put a lid on the bucket and went back to sleep. The next morning I let the shrew out in the same direction as Slithering's exit and mentally informed it about last night's dinner. I hoped it would have got its meal later.

One of the most memorable days at the farm was when I spotted the mating dance of two rat snakes. One day, walking along the fence, a faint rustling sound got my attention. I froze and stared intensely into the undergrowth and saw the head of a snake. Within minutes the head reared up and started swaying. To my surprise another head reared up to join the swaying. Both the snakes were dancing to some tune that was playing in their heads. The male kept thrusting upwards while the mate tried to keep pace with the rhythm. At one point they stretched up almost two feet in the air. I stood there as if in a trance watching them move gracefully and in tandem. Saroj Khan, the famous choreographer, could take some lessons from these two. I had no idea how long I stood transfixed watching the dance when a loud thud broke the spell.

Out in the open the two entwined bodies of the snakes lay a few feet from me, glistening in the bright sun. They were huge snakes and together as one they looked thicker and formidable. With my trance broken, I rushed back to get my mobile and took a couple of pictures. A few more minutes of swaying to the unheard number and the deed was done and they separated. The larger male slithered away while the female just lay there presumably exhausted. After a while it went in the opposite direction.

I was sure that the next monsoon we would get to see baby rat snakes and I could tell them that I had watched their conception, that is, if I learnt to speak their language by then.

Another equally amazing event was the spotting of a moth. It did not seem so dramatic till much later when we found out which moth it was. I was pottering around the banana trees when there was a flutter and a huge orange moth flew past me. I chased it to get a better look (no camera). It was real big but I could not identify it. Later I described the moth to Meena who did a check on the Internet. It was none other than the Atlas Moth, the biggest in this part of the world. It is sad that I did not have my camera to capture this sensational creature.

One thing was certain—our organic ways had brought back a lot of these creatures to the farm. No fertilizers and chemicals meant that they found it conducive to survive in our environment. The hares, the moths, the shrew and Slithering were all back. We are happy to have them back, except maybe the house guests. Some may call us antisocial. So be it.

The Dreaded Event

On the fateful day of 14 August 2014, destiny decided to test us. The dreaded event of a snakebite finally happened at the farm.

Baban and I were cleaning the fence of creepers and sundry other stuff when we decided to take a break for a cup of tea. As we walked back to the house he went to tie his bull to another chikoo tree. Just as I reached the back porch of the house, I heard a painful scream from Baban and the next minute he was running towards the house yelling that something had bitten him. He was sweating profusely. I calmed him down, gave him a glass of water and ran towards the chikoo tree to see what had bitten him. If it was a snake it would be better to identify it. Unfortunately I was too late and the reptile had slithered away in panic like Baban. I walked back gingerly to the house to attend to the victim. A closer look at the wound site with a pair of magnifying lenses and it was clear that there were two bite marks on the second toe of his right foot. I measured the distance between the marks and it was around 2.7 millimetres. That surely was a big mouth and had to be a reptile. There was an amber coloured liquid mixed with blood that was oozing out. Luckily the depth of the wound was not much and it looked like the fangs had not sunk into the finger but just grazed it.

I washed the wound with Dettol and gently squeezed the blood out of the bites. Though we had not seen the snake and had no idea if it was venomous or not, we decided not to take any chances. We left for Kasa hospital which would surely have the antivenom. I informed his family that we were on our way and they could join us at the hospital. There was no point in going to the village since experience had taught me that they would delay the whole plan of going to the hospital.

By the time we reached the hospital his foot had swollen and the toe was turning into a nice purple colour. The doctor as usual asked if we had seen the snake, which we hadn't, so he recommended a blood test to check if the anticoagulant had started working. The test came out negative, probably because we had reached there quite fast and the poison had not yet spread.

Anyway the doctor said Baban would have to be admitted to the hospital and since the snake had not yet been identified, we would have to monitor him and watch out for symptoms of poisoning. I tried to convince him to go ahead and administer the antivenom but he disagreed saying that it was like poison and unless the victim showed other symptoms it might not be a good idea. They started the saline drip and pumped some antibiotics into it. We were asked to monitor him and watch out for symptoms like paralysis, drooping eyelids and bleeding gums or ears and inform the doctor. His family had reached by then with almost half the village in tow. With so many people around, I told them what to look out for and sneaked back to the farm and grabbed my manual on reptiles by J.C. Daniel.

After reading about all the snakebites and their symptoms, it looked like it was a Russell's viper that had snapped at him. After a quick lunch, with Daniel under my armpit for reference, I went back to the hospital.

By afternoon there were no other symptoms except for abdominal cramps and loose motions. À la Ron Weasley of the Harry Potter series I would go into the doctor's cabin after each bout and ask, 'Is it time to panic?' He would nod wisely at me and say, 'Not yet.' By now Baban's foot had become huge and it was extremely painful too. I requested the doctor to help out and he administered a Diclofenac injection which gave him some relief.

Each hour passed and it looked like ages before it was 10 p.m. He was surely out of danger since twelve hours had passed and no other symptoms were showing up. The cramps and motions continued till the next day but petered out by afternoon the day after. They kept him for one more day and released him when the cramps stopped and he was able to consume solid food. We were sent back with some medicines and asked to return after three days.

Once back at the village there was a steady stream of visitors from all over who had their own ideas on how to treat the snakebite. Finally, even though it was established that there was no poison in his body, due to public pressure we decided to go to the nearby village to an Adivasi who cured people with snakebites. We reached the hamlet of Somta where this young man came out with a bunch of dried leaves which he mixed with water and gave Baban. Then he pushed off into the jungle to return a hour later with the bark of some tree which he made into a paste and asked us to apply the same on the wound. I asked him who had taught him this and he said it was his father. On inquiring who had taught his father he explained that it was an old Parsee who had lived there many years ago.

It took almost fifteen days for the swelling to reduce. We were grateful that the episode had a safe and happy ending. The dreaded event had not left its scar on us but just warned us to be more careful when we walked around the farm.

One of the first things I did when I returned to Mumbai was to buy two pairs of rubber gumboots. They would at least give us some protection when we tramped around on the grass. I also got in touch with the panchayat office and managed to get a grass cutter from them. It is a useful machine and makes grass cutting an easy task, ensuring that the undergrowth is always trimmed.

Allahrakha

One day Baban and I decided to clear the undergrowth near the compound, armed with sickles, axes and sticks. Around 5 p.m. there was a movement in the grass ahead that caught my eye. I held Baban's hand and pointed at the pale brown object a few feet ahead. '*Sasa*,' he whispered. Sasa means hare and is a delicacy here. He slowly parted the grass and instead of the two long ears we expected to see we saw two big eyes and a white round face looking at us. It was an owl!

Grabbing it by the neck we took it to the house for a closer look. It was a baby and she could not fly at all. She probably fell off the nest and was trying to hide. After a good photo session and a quick look in Salim Ali's book on Indian birds, I identified it as the common barn owl. Though the name says 'common', the spotted owl is seen more than the barn owl at the farm.

Now we had to do something with her. I was not sure of hosting her till she got better. A few calls here and there and finally I got through to a forest official who promised to send someone to pick her up.

That done, we had to be good hosts to our new guest. What would she like to eat? After a consultation with Meena we decided it was worms. Soon we were on all fours searching for the elusive worm. Finally I pulled one out and of all the worms in the world it turned out to be an earthworm. Here was a moral dilemma: should I sacrifice a friend of the farmer to feed a bird? Finally statistics solved it. There were far more earthworms than barn owls. I held it to her but she did not accept it. Baban said, 'You know, in the wild, the mother feeds the young from her beak.' I glared at him and said that I refuse to hold the worm in my mouth to feed my guest. She will have to learn table manners if she eats here. The best I could do was chop the worm into small pieces for her. Anyway to give her some choice I kept one small piece of idli (in case she was a south Indian) and some chopped boiled egg too. She was kept under a basket with a heavy stone on top so the dogs didn't get her. This done, Baban left for home and I went about my evening chores.

Around 9 p.m. I got a call from a slurring Range Forest Officer (RFO) asking for directions. I knew they would never find the place so I asked them to reach the village and pick up Baban who would guide them to the farm. Four guys landed at the farm smelling like an arrack shop. The RFO examined the owl, checked each wing and declared that it was a barn owl. I just casually remarked that Salim Ali had confirmed it. My eyes popped out when he looked up and asked, '*Ekde ale hote ka?* (Did he come here?).' In a state of shock I just replied, '*Nahi*, WhatsApp *kela* (No, I just WhatsApped him).'

Once his IQ had been established he started his paperwork. Reams of papers had to be filled. Questions ranged from why I had decided to clean that section today to if I had seen the nest by any chance. As they went about this inane task I noticed the driver standing far away in the darkness. I called him to come and sit on the porch but he refused to budge. So I walked up to him to figure out what the matter was. He explained, 'I am a Hindu Brahmin and we do not see an owl's eyes after dark.' He explained, 'The owl is the carrier of black magic and if you look into its eyes the curse is on you.' No amount of convincing that the baby's beautiful eyes carried no curse made him come see the owl.

Finally, after an hour, the papers were filled and signed. We packed the owl in a cardboard box and I was glad to see that she had eaten the earthworm . . . may its soul rest in peace.

While the paperwork was being done our drunk RFO took pictures of himself and the owl in different poses. He grandly announced, 'This is for the media release tomorrow.' He then went on to say that he had seen a movie of Amitabh Bachchan once where the actor had the same bird as his constant companion. I replied, 'Sir, I have also seen the movie. It's *Coolie* and the bird he carries with him is a falcon and it's called "Allahrakha" in the movie.' Nothing registered except the name and he told me, 'It's a nice name. We will call the owl Allahrakha.'

It was almost 10.30 p.m. when they left after assuring me that the owl would be taken care of.

After they were gone I looked at the pictures of the owl I had taken and wondered if she would one day soar the night sky and maybe pay me a visit.

Allahrakha is welcome.

Hen Log

Ever since I got the farm, I had always wanted to keep hens. A couple of years after we had settled in I broached the topic with Baban. He said it was not a bad idea, except that each time I went to Mumbai for the weekend there would be a chicken party in the village. He said, 'People know when you leave and the fence is not exactly foolproof.' I thought he made sense and put the idea on the back-burner.

In 2013, Meena was posted to Pakistan as *The Hindu's* correspondent there. Now, I had no need to go back to

Mumbai for the weekend and thought the time had come to start our poultry farm. Baban agreed with me and offered one of his hens. The deal was that once she multiplied I would return a hen to him. It sounded like a great deal and I instantly accepted his offer.

The next morning he came with a beautiful hen from his lot at home. She soon settled down at the farm. I placed a small box in the front porch for her to roost at night. We christened her 'Mother' since it was expected that one day she would have her own brood. She was big enough to start laying eggs and a week later she started. It was fascinating to watch her go around searching for the ideal place to lay her eggs. She made peculiar noises as she went about looking for the place. Baban told me that the crown would turn crimson red when she was about to start her egg-laying cycle.

Her first lot was twelve eggs, one on each day. It was a joy to go out in the morning and pick up the warm egg she had laid. I learnt that hens always laid one egg at the same place and at the same time every day. They also did not know how to count and it sufficed if you kept one egg for her to identify the place she had laid them. Baban explained that if there was no egg she would conclude that the place was not safe and start looking for another one.

I was keen on multiplying the lot, but there was no rooster to mate with Mother. These eggs were unfertilized and there was no point in letting her hatch them. They would just spoil after a few days. We wondered how we could get a rooster since Baban had only one at home which he could not spare.

As luck would have it, one fine morning, I saw a magnificent rooster saunter into the farm courting Mother. We had no idea where he came from or who the owners were. For a few days he would come in the morning and leave the farm in the evening. Then he just decided to stay on. He never sat in the basket where Mother roosted but perched himself on a nearby tree. Baban kept wondering whose rooster it was since no one had reported a missing rooster and it was not normal for them to leave their homes and come to a strange place.

The next lot of eggs was surely fertilized and we let Mother hatch them. She sat for twenty-one days on the eggs, venturing out of the box only once a day after the sun was out. She would step out of the box and stretch her legs one at a time and then fluff her feathers and run from the house to the gate. On her way back she would peck here and there, eat the food I would keep on a tray for her, drink a bit of water from the bowl and go back to the box. Once inside she would carefully turn the eggs over, ensuring that each one got the warmth of her body.

After twenty-one days, one morning, I heard tiny cheeps coming from the box. I peered in and saw the most beautiful chickens I had ever seen. They were just tiny balls of golden feathers with a small beak and two wobbly legs. A couple of more days and Mother had hatched seven of the eggs.

Once they were able to stand on their legs Mother took them out of the box and into the area around the house. I would watch them for hours as they went about following their mother and trying to imitate what she did. They were so tiny that at times a couple of them would get left behind as they tried to keep pace with their mother. Mother was extremely careful and would always call out to the ones who were left behind, even going back sometimes to bring them along with the rest.

One afternoon, I heard her clucking frantically and I rushed to check what had happened. I reached her just in time to see a crow pheasant soar away with one of the chicks in its claws. The chicks were too small to run away and were excellent prey for the crow pheasants. Within a week the entire brood had been preyed on either by the crow pheasant or the shikra. I felt I had failed Mother in some way.

When she had the next brood we were careful. We made a small cage out of an old wooden crate and ensured that the chicks did not get out till they were at least a month old. By then their feet were strong and they could scamper away when Mother gave them alarm calls on sighting birds of prey. This worked and after one month we had four healthy chicken. The brood this time was smaller and two died after being hatched.

In 2014, Meena returned to India from Pakistan after being expelled from there and was posted to Delhi for her next assignment. I went to Delhi to settle her in the new house when one morning Baban called frantically. He said, 'All the chickens are dead.' It seemed like a mongoose attack. Three of them were lying on the bloodied front porch with their throats slit and one had been dragged to the edge of the farm where the remains of the carcass were spotted. I had failed Mother again.

On my return to the farm, I called the local welder and designed a cage with netting from an unused gate that was lying at the farm. We decided that at night the chickens would have to be moved to the cage and let out only in the morning. It was common knowledge that the mongoose attacked only at night.

Mother's third brood had four chickens and the cage worked. They soon grew big and we had three hens and one rooster. We named one Mother Junior for she looked like a replica of Mother and the other two Turkey and Ostrich after their looks. The rooster was Chief for that's how he behaved. Chief turned into a magnificent rooster and was extremely friendly, even eating out of my hand at times when I fed him.

In a few months, Ostrich and Turkey had their own brood. There were so many of them now that it was difficult to name them. Meena decided to call them Hen Log, after the famous television soap opera *Hum Log*. Sometimes we imitated veteran Bollywood star Ashok Kumar who appeared after each episode and summarized it in his characteristic drawl. I kept my part of the deal with Baban and returned a hen and a rooster as bonus to him.

A few months later, in 2015, Meena quit her job and moved to the farm. Finally, both of us lived at the farm. There was no need to go back to the city. After eleven years, we were on track as per the original plan that we had started out with.

The Cats and Pepper

After she moved to the farm, Meena wanted to keep a few cats. She loved cats and at one point in their tiny flat in Mumbai they had seventeen cats. We thought we would start with a pair at first. I contacted Vithal and asked him to get us a pair of cats if he found any.

One night, as we were preparing to sleep, we heard calls from the gate, 'Seth! Seth!' I took the torchlight and went to the gate to find Vithal with Baban in tow and a small sack in his hand. It was the kittens. It was late and Vithal was pretty drunk. One always knew when Vithal got drunk as he would switch over to speaking in Hindi after having a couple of drinks. He said they were a pair—one male and one female—and he had found them in Nanivali. We kept the kittens covered under a basket for the night.

The next morning we realized that what he had got us was a pair of male cats. They were tiny and cute and anyway there was no way of returning them. We named them Crash and Eddie after the two possums in the animation movie *Ice Age*.

The pair, after they got over the initial fright, were fun to watch as they romped around the farm. We got fish from Nanivali or Kasa whenever it was possible and they just loved it. We set the rules from day one for them. They were not allowed inside the house. Food was served only at scheduled times and they would have to wait for it. Crash learnt the rules quite fast and obeyed them, while Eddie would try and sneak into the house and steal some food if possible.

Once the cats had settled down, we thought of getting a dog. Both of us loved dogs and had kept them as pets earlier. We spread the word around and got many an offer which we refused for some reason or the other. It was on one of our trips to Mumbai, when we had stopped for breakfast at a restaurant, that we ran into Meena's old friend Mini and her husband, Zak. She and her husband were staying in Dahanu and ran a kennel for dogs. Zak knew a lot of people and said that he had heard of a pup which someone wanted to give away. He promised to check on it and get back. The next day we got a call from him saying that the pup was still available in nearby Gholvad.

We went to Gholvad and met the owner. He had got the puppy from Mumbai but his family was not ready to accept dogs and he was looking for someone to take the puppy and give it a decent home. We went to see the puppy who was tied in the back porch. It was only two months old and looked like a Labrador. She was completely black expect for one white patch on her chest and one paw. We just loved her. She sat quietly on Meena's lap in the car for the hour-long journey back to the farm.

We asked people to suggest names for her and finally decided on Pepper. Pepper grew to be a friendly dog and even Baban, who had a mortal fear of dogs, started liking her. After she settled in at the farm, one of the first things she decided was that she hated the hens. Especially after one of them tried to steal some of her food, she just chased them when she got the chance.

We had to think of a solution quickly. There had been a couple of instances where we had to rescue the hens from her jaws. Each time there was an incident, Chief the protector of the brood would attack Pepper. He lost quite a few of his beautiful plumes during these skirmishes. Finally, we decided on a schedule. The hens would remain in the cages till 9 a.m. and were released only after Pepper was tied up in the front porch. They were free till 3 p.m. when they would have to return to their cages and Pepper would be released. It worked beautifully since at no point were both the hens and Pepper free together.

Once, when Meena went for a walk with Pepper on a leash Chief attacked Pepper. Meena shooed him away and even threw some stones at him while she held Pepper on a tight leash. We presume that this incident was what made Meena enemy number one in Chief's eyes. Every time he saw Meena he would attack her. When she told me this I did not believe her. I thought she was imagining the whole thing. She then stepped out into the back porch and said, 'See, he will appear now.' Sure enough, within thirty seconds, Chief appeared out of nowhere and started attacking her legs and causing blood to spurt out. He had a sharp beak and would ruffle his feathers into a circle around his neck when he attacked. No amount of shooing or throwing things at him worked and he continued his attacks on the door even after she had shut it.

In the village, people were surprised to hear this story. They had never heard of attacking roosters. We soon realized that it was not only Meena that Chief attacked but any woman who came to the farm. Meena's mother, my mother and our friends were all victims of Chief's attacks. One of our friends, Romilla, has a deep scar on her thigh as a memory of Chief's attack. Things were getting out of hand and one day, as she was leaving for Mumbai, Meena said, 'Ravi, you have to decide. It is either me or Chief at the farm.'

That evening, I called Sagar, Baburao's son who ran a chicken shop, and asked if he wanted a rooster. He was excited since the locals loved the desi *kombda* or rooster and they fetched a high price. The next morning Baban helped me tie up Chief and I took him to the shop. Sagar offered to pay me but I declined. I did not feel like taking money for Chief. I gave him one last look before leaving the shop. Chief was regrettably sacrificed and peace was finally back at the farm and people could walk around without the fear of being attacked.

One of Meena's students from the college where she now teaches once a week called one day to say that they had found a small kitten abandoned near the college and had no idea what to do with it. She asked if we would like to take the kitten and give it a home. Minimus was a small white ball of fur when we got her to the farm. She must have been just a few months old. Mini soon became the princess of the farm. She got the best food and all the rules were bent for her. She had free access to the house, sometimes even spending the night on our bed.

Mini behaved like a dog at times, following me all around the farm as we went about our work. She got along well with Crash and Eddie and played with them. Though Pepper hated the cats and always chased them, she gave a free hand to Mini. At times we found both Pepper and Mini sleeping on chairs next to each other.

We had always feared that the tomcats would leave us and that is exactly what happened. The first to leave was Crash, followed by Eddie a few weeks later. We searched for them all over the farm and even asked around in the village but they were never seen again. Meena kept saying that it was time to get Mini spayed, else we would have a house full of kittens if she got pregnant. We had already got Pepper spayed a few months ago. For some reason or the other it did not happen and soon Mini got pregnant. We had no idea who the father was for we never saw any male cats around.

Mini decided to litter on the book shelf in the spare bedroom. The room was converted into a maternity ward for the five kittens that she delivered. She gave birth to one female and four male cats. We could not keep so many cats and asked Baban to check if anyone wanted them in the village. Baban agreed to keep the female and our friend Fawzan took two males. We were left with two males whom we named Dada and Whitey.

Mini had one more litter of six kittens before we took her to the animal hospital in Mumbai and got her spayed. The second litter of three males and three females was grabbed by my friends Vipul and Purvita and one pair went to Atmaram in the village.

We had no idea that the demand for kittens was so high in the village. Months after we got Mini spayed there were many who stopped me on the way and asked if I had any spare kittens. I told them that I would not be having any but we had hopes that one day the cat at Baban's would litter and they could have some from there.

Unlike the tomcats Mini turned out to be a master hunter. After her first litter was big enough to eat, she started hunting and would bring home rats for the kittens. On some days she got home four rats one after the other. We had no idea we had such a large stock of rats on the farm. We were grateful to her for reducing the rat population on the farm and could soon see the results. The number of pumpkins eaten by the rats had dropped drastically. I still remember one particular year when they ate up eight of the pumpkins that had grown at the farm.

She also had a liking for the buff striped keelback, a thin snake that she would catch and play with. She never killed it and let it go after playing with it for a few hours. Each time she got one of the snakes home, Meena would say, 'I hope she knows the difference between a keelback and a viper.' Her other favourite was the skink, which was a regular victim of her hunting skills. Besides the outdoor hunting she also specialized in catching lizards at home. Soon, most of the lizards had disappeared and the few that were left inside the house went about minus their tail which they dropped when attacked.

Sometimes at home, Mini would design her own games to play. A bottle cap, a clip fallen on the ground, the rug near the door or anything that caught her fancy became the object of her games. She would let out low growls and leap at the object, throw it in the air and try to catch it. If it was a rug that caught her fancy she would crawl under it and emit gurgling noises before sneaking out. When she is in this mood, we call her VP Nintendo Corp for the games she invents.

13

Village Economics and the Man Who Hates Banks

Economy in the city meant keeping a watch on the Gross Domestic Product (GDP), inflation, our industrial growth rate and the Reserve Bank of India (RBI) repo rates. Here in the village it was just simple economics. Do you have money to survive? Most of the village was dependant on agriculture when I moved here in 2004. Everyone had children who were either in school or college. The parents worked the land all year long and just about made money to survive.

Decades ago the village was completely dependent on the monsoon for agriculture till the Surya canal project got them water for a second crop. The land holdings kept dropping as each generation distributed the land amongst themselves. The land is now completely fragmented. Everyone has pieces of land here and there. No one has contiguous pieces of land so they can think of investing in a fence to ward of cattle. If they had a fence they could

Village Economics and the Man Who Hates Banks

grow vegetables and fruit trees which could augment their income.

They grow rice and some pulses in the monsoon. The canal water is used to grow groundnuts for oil or rice again. Other than that they have a few vegetables that they grow around their houses. For the rest of their existence they need the elusive cash.

Cash usually comes from the sale of paddy straw or rice if they have excess. The paddy straw market is like the stock exchange, going up on demand and crashing on excess supply. Some years you can get Rs 1500 for 500 kilograms and sometimes it crashes to Rs 900 for 500 kilograms. Though the government has announced a Minimum Support Price (MSP) for rice the broker who comes to the village rarely gives that price. Rice is usually sold at around Rs 10 a kilo or maximum Rs 12 if the quality is good.

Some of the enterprising villagers grow vegetables after the monsoon which they cart to Boisar or Kasa to sell. Depending on the yield they earn about Rs 20–25,000 every season.

Most of the children study till class VIII in the nearest school. A few interested in studying further go to Nagzari 12 kilometres away and study till class XII. The school in Nagzari just started a graduation course a couple of years ago. Until then the nearest college was in Dahanu or Palghar. The youngsters have no interest in farming and are keen on working for companies. Luckily for our village there is the Maharashtra Industrial Development Corporation (MIDC) industrial area at Boisar that has hundreds of factories that need labour.

Most of them work either in the manufacturing units or the packing units. A few work in some readymade garment factories that have sprung up in Boisar. These kids earn about Rs 10,000 to start with and if they manage to stick on they get regular increments and bonuses too.

The first thing they bought when they got their salary was a fancy smartphone and the next plan was to save for the down payment of a motorcycle. Once they got drawn into the debt cycle there was no looking back. As one EMI got over they would start the next one. It could be a bigger television or a fancier mobile or the latest motorcycle. They hardly contributed to the house expenses and one could see their parents still struggling to run the house on the meagre income they generated.

Besides these expenses, there was also the infamous wedding celebrations for which they would save. Each one wanted to outdo the earlier wedding. The flashier the wedding the better and the occasion would be discussed in detail for days on end.

Umesh, Anantha Kaka's son, studied up to class VIII and landed a job in Pantaloon, a readymade garment factory in Boisar. He was part of the assembly line which manufactured pants for men. He worked hard and in due course was promoted to line supervisor. He then decided he wanted to get married. Over the years he had saved some money but that would not suffice for a grand wedding. His plan was to have a band for at least two days and serve the best mutton for his wedding.

With no solution in sight for the money, he did the next best thing he could think of. He quit his job. I was shocked at what he did, till he calmly explained the reason to me. He said, 'How else will I get my provident fund?'

He waited out the two-month cooling period to apply for withdrawal of the provident fund. He got the money and a month later had his dream wedding. Once the euphoria of the wedding had subsided he just went back to the company and rejoined. He was recruited back on the assembly line at a lower salary than what he had drawn earlier. He was confident that in a few years he would be back as line supervisor.

A year later he was promoted and has now moved to Boisar with his wife and kid to live in a rented apartment close to his place of work. He visits the village on holidays or if there is a death in the village. His old parents continue to till the land and live in their hut in the village.

Pavan Kaka is a person that one could study to understand the village economy. He lives with his wife in the village while his only son stays with his wife in her village in Saphale close by. They run a garment shop in Manor and for setting it up they had borrowed money from Pavan Kaka. The boy never gives any money to his parents. Pavan Kaka survives on selling his paddy straw and some of the rice he grows.

Though the son never gives a single paisa to his parents, each time he visits the village he would return with loads of stuff. A tin of groundnut oil, a sack of rice, a bag of vegetables and sometimes a can of kerosene too. I saw this once and asked Kaka if his son paid him for all this. Kaka said, 'He did offer, but I did not take.' I knew it was a lie to protect his son's reputation.

Each year before the monsoon, he borrows money from the credit society to buy seeds and fertilizers for his rice. After the paddy is cut and threshed he tries to repay the money by selling the paddy straw. If there is a drop in production for some reason he is left with less money after paying off his debt. It is a balancing act that he performs every year. Both of them are growing old and finding it difficult to till the land. I wonder what will happen when it becomes impossible for them to do so. He is over sixty-five years old and applied for the government pension scheme three years ago. Every year he makes a trip to Dahanu to fill up the forms. He went this year too and is hopeful that it will be cleared and he will start getting the pension.

Most of the families in the village have a similar tale to tell. There are some of them like Dashrath and Prakash who have stopped tilling the land and are working as security guards in Boisar. Arvind moved to the city and is driving an auto there. They at least get some money home at the end of the month.

There are a few in the village who try to generate some additional income. One villager has set up a rice mill in the village by taking a bank loan. Another has set up a bottled water plant in the village. Dina has got two trucks which he uses to deliver construction material. Hari too has a truck and is into the construction material business. Baban's eldest son Jayesh has a Tata Magic which operates locally and a

Tavera for long distances. Baban's younger son Mahesh has completed a basic computer course and is working in a factory as a production assistant.

When I came to the village in 2004, I was the only one with a four-wheeler. Now, the village has quite a few vehicles. Dina was the first to get a Maruti Swift, followed by Jairam, Hari, Subash, Vikas and Baban's son Jayesh. All the people who got the vehicles are doing some business other than agriculture.

I realized that with such low land holdings, the income generated from the land would not suffice for them. The families were also growing and they had to look for an alternative source of income to survive.

The Man Who Hates Banks

Damu is the brother of Lahu and Sridhar Kaka, my nearest neighbours in the village. He was an extremely hot-tempered man and used to get into fights with almost everyone in the village. Some thirty-odd years ago, Lahu and Damu had a huge argument over some family matter and in a fit of anger Lahu Kaka hit him on the head with a stout stick. A bleeding Damu was rushed to the hospital. After his recovery in the hospital he vowed never to return to the village and left for a small fishing village—Kharekuran.

He worked on the fishing boats there, never once coming back to the village. Suddenly in 2013, the villagers saw an old balding man with a huge white beard, a stout stick in hand

and a small plastic bag enter the village. He went straight to Mohan's house and sat on the porch. Everyone gathered around and someone shouted, 'It's Damu. Damu is back!' The young angry man who had left the village had finally returned a bald old man.

Damu returned to his land which is a few metres from our gate. He did not speak to anyone in the village and when I tried to strike up a conversation he just nodded. He got bricks from the nearest kiln, paid in cash for it and built his house all by himself. It was just one room with a couple of asbestos on top for a roof. A small wooden plank was the door.

I passed by his house every day, but he never uttered a word. He got up in the morning and left to do odd jobs for people in the nearby village. He did not work for anyone in the village. I once asked if he would like to help at the farm as I needed an additional pair of hands. He looked at me and said, 'You have Baban.'

Damu always carried a stick with him. He also had a small bundle wrapped in plastic that he kept under his armpit. He was never seen without it. Even when he went to the river to bathe the bundle would always be with him. There was much speculation on what exactly it contained. It was someone in the village who revealed that it was his life savings, hard cash that he carried on him. It was rumoured that it was around a lakh.

Damu would get up in the morning, go to the river and leave for work by 7.30 a.m. He only returned for lunch at noon, which he made himself after coming back. He left for work again at 2 p.m. After his day's work was done he would collect his daily wages of Rs 150 and head for Dhamatne, the next village 3 kilometres away. There he would have a couple of glasses of toddy before walking back home. On the way he would pick up some chicken or eggs or a couple of vegetables for his evening meal. He cooked his dinner and went to sleep. This schedule was common knowledge and he always adhered to it.

I met some of the elders in the village one day and broached the subject of Damu. I told them that now that it was common knowledge that Damu carried cash with him, he was at risk. He stayed far from the village in a hut with only a plank for a door. He was a sitting duck. What if they harmed him before taking his money? They said, 'We have asked him many times to open a bank account, but he does not listen.'

I met Damu that evening and asked him if he needed any help to open a bank account. He just nodded his head. I said, 'Kaka, I can take you by car to the bank and help you fill the forms.' He replied, 'They will never give you the money.' At first I could not get the point. I asked, 'Who will not give you?' He said, 'The bank.' It was the end of our conversation. I spent the next ten minutes explaining the risk of staying alone with so much money which is common knowledge in the village. My monologue was just met with silence.

The next day, I went back to him and said, 'Kaka, if you don't like banks we can open an account in the post office.

There is one at Tawa.' He replied, 'They are the same.' I just gave up.

A year later our fears came true. One afternoon, we heard some shouts from near the gate and Baban and I went to check. We saw Damu running up and down screaming, '*Chor! Chor!* (Thief! Thief!).' Damu had come home in the afternoon as usual, entered the house and hung up his precious plastic bundle on a nail near the door. He then proceeded to cook his meal. He claimed to have stepped out of the house for just a minute to get some wood from the heap next to the house. He only realized he had been robbed when he went to pick up the bundle before leaving for work at 2. It was obvious that someone had been watching his routine and had sneaked in at the opportune moment to steal the bundle.

We went down to the river armed with sticks and a while later found the plastic bag in the bushes. It was empty except for his election card. There was nothing around. I asked how much was in the bag. He muttered, 'Rs 70,000.' I tried to convince Damu to accompany me to the police station to lodge a complaint. He would not agree and asked us to go away. We quietly left to let him mourn his loss alone.

Many people from the village went to console him and some even tried to give him money. They all told him to lodge a complaint but he refused to listen to anyone.

A few months later I spotted Damu on the road with his stick. As I drove past him I noticed the plastic bundle had returned under his armpit though it was much smaller and thinner than the earlier one. It was obvious that the theft had not changed his hatred for the banks.

The Demon of Demonetization

One of the main differences between the city and the village is the complete lack of banking activities in the village. Everything and everybody dealt in cash. They all had bank accounts but no one used them unless they wished to deposit some large sum which anyway was not normal. Each time they received money, be it the loan from the society or some compensation from the government or their pension, they would all rush to the bank and withdraw it, keeping only the minimum required to maintain the account. A few kids who worked in big companies flashed their debit cards and spoke at length on how they could withdraw money from a machine while everyone listened to them in awe.

I had suggested to Baban that I would transfer his salary to his bank account each month so he got it even if I was not around or did not have the cash to pay him. The nearest ATM is in Boisar and sometimes it runs out of money. He was not comfortable with the suggestion and refused it. He said, 'I will have to go to the bank each time I need money and besides the expense of going to Kasa, I will also be wasting a day waiting in the serpentine queues.' So it is cash payment that he preferred even if it came a few days late.

When the demonetization of Rs 500 and Rs 1000 notes was announced there was panic in the village. A few elders who could not understand what was happening felt that they had lost all their money. It took a while to explain to them that it was not true and they would get their money exchanged.

Next morning a large crowd gathered at the sub-post office in Kasa. I was part of the crowd as I had some old Rs 500 notes that had to be exchanged. The queue outside the only bank in Kasa was more than a mile long. Someone had informed us that the post office was better as only a few people were aware that the exchange could be done there.

In one corner of the ground of the post office I saw a woman huddled up and crying. I went up to her and asked if I could help. She nodded and showed me her life savings of four Rs 500 notes and a tattered copy of her Aadhaar card. Gajari Anya Dhangad had come from the remote hamlet of Dongripada and had no idea of the upheaval caused by demonetization in the country. Next to her, shivering with fever, was her daughter-in-law, whom she had brought to Kasa to be treated. It was only when the doctors in Kasa refused to take her money that she knew something had to be done.

I helped her fill up the form and submit it at the counter. It was almost 2 p.m. when the postmaster arrived with a bundle of new notes. He agreed to a request that Gajari be given the money first as she had to go to the hospital. She clutched her new Rs 2000 note and left the post office.

The village of Kasa had two banks, the Bank of Maharashtra and the Thane Co-operative Bank. But the latter had closed operations. This meant more crowds at the sub-post office and the only other bank. People spent the entire day waiting for money, only to be told to return the next day.

Not everyone was as lucky as Gajari to get their money exchanged. Ladkya, a daily-wage construction labourer who was among those waiting in line, looked dejected when the postmaster announced that they had run out of money. Spending the day at the post office meant he had lost that day's daily wage and now faced the prospect of losing a second day's income too. He had no money to go back to his village, about 10 kilometres from Kasa. He prepared himself for the long walk back home.

As Ladkya walked into the sunset, one of the persons hanging around at the post office remarked in Hindi, '*Magarmach ko pakadne ke liye talab khali karen, toh sab machli khatam ho jayenga* (If you drain the lake to catch a crocodile, all the fish will be finished).'

With no money in the banks and establishments refusing to take old notes, the villagers were worried about the crops they had sown for the rabi season. They could not buy fertilizers or pesticides and the unexpected cold wave in Maharashtra had caused quite a few pests to appear. Now they faced a bleak season ahead if things did not improve soon.

While the farmers were fretting over their crops and daily existence, the businessmen too were at their wits' end. Sandeep Pawde, who had a roaring brick kiln business, was worried as no builder was buying his material. He even offered them credit but they were least interested. He wondered what would happen to the money he had invested in making the bricks.

Jinendra, the guy who bought paddy straw from the villagers, had more than Rs 3 lakh in old denominations and the paddy season was about to start. He had no idea how he could exchange so much money and pay the villagers who had already started getting the straw.

I spotted Sitaram of our village nearby and waved to him. He came running and almost fell at my feet. He had come to Kasa to get his money exchanged and now had only the Rs 2000 note with him. No one was willing to give him change and he didn't know how to return to the village. I offered him a lift back.

As we left the post office after exchanging my old notes, I spotted Gajari and her daughter-in-law sitting by the road side. She was in despair—no doctor was willing to give her change for Rs 2000. The local grocer was willing to give her change provided she bought provisions for at least Rs 500, which she did not want or need. With no solution that one could think of, I gave her Rs 50 and guided her to the nearest government hospital where they would at least treat the poor woman. I thought of what Damu had said the other day, 'They will never give you the money.' It was as if his prophesy had come true.

14

Market Initiatives

Lessons in Supply Chain

The first year at the farm, I planted ladyfinger (bhindi) in the entire backyard. They grew well and soon the harvest was more than what we could eat. I started looking around to see if I could find a market for them. I heard that there was a vegetable vendor who came every evening to the village and bought any kind of produce which he carted to Boisar in his tempo.

That evening I went and stood at the corner of the road to try and catch him. I met Hari Bhai, the vendor, who greeted me profusely as if I was his long lost friend. I asked him if he would like to buy my bhindi. I had already learnt a lesson and did not even mention that it was organic. He replied, 'Why not? I am at your service.' I asked how much he would pay. He said Rs 4 per kilo.

My jaw dropped. I had just checked the rates in Mumbai for bhindi and they were selling in Goregaon outside our house for Rs 20 per kilo. I told him, 'You are looting the villagers. I just checked and it is Rs 20 in Mumbai.' He was very calm and said, 'Looks like you are selling for the first time.'

He said he would explain to me why it was Rs 4 only. He said, 'I come all the way in this vehicle from Boisar to collect the vegetables. There is a cost to it. Let's put it as Re 1 per kilo. I have a family to support and need to make some money at the end of the day, say Rs 2 per kilo. I sell this load to the Boisar wholesale vendor for Rs 7. He collects from many like me and sends it by train to Borivali in Mumbai. There is a cost of transportation, either a ticket or a bribe to the ticket checker. Let's say Re 1 per kilo.'

Before he said it, I remarked, 'Of course, he too has a family to support.' He smiled and said, 'Now you are getting it. The Borivali vendor then distributes the produce and it goes to the Goregaon municipal market. Whether it goes by road or train, there is a cost involved, say Re 1 per kilo. He adds his share of Rs 2 per kilo. The vendor outside your building then takes the produce from the market to his stall. He has to bribe the municipal authorities and police to set up a stall on the road. Let's say Re 1 per kilo.'

He continued, 'We have still not accounted for losses in transit and spoilages. You are educated and you can calculate what the final figure is and why Rs 4 is a good price for you.' I did a quick mental calculation and it came to Rs 18 by the time the produce reached the stall outside our house in Goregaon. This of course included the profit at each level. I was so numbed by the calculation that I did not even bother to ask him about the balance amount.

Before starting his vehicle he gave me one look and said, 'You are a good man and from the city. For you I shall give Rs 5 per kilo, but don't tell anyone.'

I returned to the farm after having been through my first lesson on supply chain in the village. The only way a farmer could make money was to break the chain and sell directly to the market. But with such low land holdings and small quantities how could one do it? I could not afford to go to Mumbai every alternate day with just 10–20 kilograms of bhindi. Even if I did where would I sell it? I realized that marketing farm produce and that too perishables was not an easy task.

The next day, I told Baban to throw all the bhindi back into the soil. I vowed never to plant more than what we could eat.

After our first experience of selling moong in the market we had already reconciled to the fact that direct sales was the best option. Hari Bhai's supply chain tutorial just confirmed it. Grains, pulses and oil had long shelf lives and were easy to cart and distribute in Mumbai.

I still had chikoos and needed a market for them. Whenever I returned to Mumbai I carried a huge bag of around twenty dozens to sell. The local fruit vendor outside our house bought them from me for Rs 15 a dozen and sold it for Rs 30 a dozen. When I confronted him about the huge difference in the price, he just smiled and told me, 'Why don't you sit next to me and sell?'

The first five years we sold our produce like rice, dal and oil to the dedicated list of friends and acquaintances who had opted to be on our customer database. We sent emails or SMS's to all of them and collected the orders. We packed the produce as per the orders and I would go and deliver them while also collecting the money at the time of delivery. It was the only option we had to beat the supply chain and earn some money. Of course the additional effort of packing and transportation was there.

The organic market has changed in the last five years. There is greater awareness and in Mumbai a number of new outlets have opened which cater to organic produce only. One of our friends, Chaitanya, also opened an outlet in the suburb of Vile Parle. He procured his stuff from various organizations but was clear that the first choice would be to buy directly from the farmer.

We tied up together and he promised to take whatever I grew. The price he paid us was marginally lower than what we would earn if we sold direct. But the arrangement suited us very much. We contacted all our customers and informed them of our decision to opt out of retail sales. We gave the number and contact details of Green Current, the shop that Chaitanya had opened.

Till date our arrangement is alive and all the farm produce goes to Green Current. Besides this, we also had the option of selling to Navdanya, another chain which sold organic produce. They had their own farm in north India and got some of the produce from there. So they wanted only certain items that did not grow on their land.

As for the chikoos, I still continue to take whatever is possible to Mumbai. The last few years we have been making very few trips to Mumbai. I have an Adivasi from the neighbouring village of Amboli who comes every week to take the chikoos. He pays Rs 8 per kilo which is low but better than nothing. Besides he and his wife pluck the chikoos, wash them and take them back. We do not have to do anything. It is better than selling to the cooperative near Dahanu which pay us Rs 5 or Rs 6 per kilo. The plucking, washing, sorting and transportation expenses also have to be borne by us.

The MOFCA Experience

After starting the transition, I made a lot of like-minded friends. There were people like Gaurang who had a computer business and shuttled between the farm and the city. Ubai was a chef who decided to move back to India from the USA and start a farm on his ancestral land near Mumbai. There was Archana, Raajen's wife, who had quit her job and moved to the farm to till the land. Vipul, who had a huge plastic manufacturing company, had sold his share in the company and bought land to do organic farming. And there was Purvita, a computer web designer who had also started the transition from the city to the village.

We were in constant touch with each other to either discuss our problems on the field or to just find out what the other person was planting. We exchanged seeds and saplings and tried to help out whenever possible.

During our many interactions and meetings we realized that besides our usual issues with power and seeds we all had the same problem with marketing. We grew a lot of stuff but had no idea where and how to sell it. Besides, since we all had land at least 100 kilometres from Mumbai city, we faced a logistics issue too.

One of our friends, Karen, had a two-acre farm near Mumbai and she had invited some people from abroad to help at the farm. As one of their initiatives they researched and prepared a white paper on the potential of organic food and its market in the city of Mumbai. At the end of the research they decided to share it with all of us.

We had a meeting of all interested people at Karen's place. The crowd was a mix of people who had farms, some shop owners, some consumers and some NGOs working in the organic field. We sat and discussed all day and in the end decided that we had to do something about the market and the gap that we all faced between the farm and the city. The conclusion at the end of the day was the same that we had arrived at. The farmer and the market had to be brought closer and the links in the supply chain broken down. That was the only way the farmer could get a fair remuneration for the efforts he put in.

It took a while for things to take shape and in 2010 we finally decided to jump into the marketing of organic vegetables. The plan was to commit consumers in the city to buy vegetables grown organically from us and to charge a deposit of Rs 500 from each consumer. This was to ensure that we had a fixed consumer base to sell rather than sit in the market and wait for consumers to come by.

Vegetables would be delivered once a week at fixed pickup points across the city. We thought of home delivery but the traffic and the logistics involved in the vast city of Mumbai did not justify it. The consumers paid the price per kilo at each pickup point.

For the production of vegetables we had our group of city farmers who would contribute. In addition we also had the support of an NGO (M.L. Dhavale Trust) which worked with a lot of farmers in the neighbouring taluk of Vikramgad to convert them into organic farmers. They also faced the same problem of marketing and were more than happy to tie up with us. For the first season we had four farmers associated with the trust who promised to grow organic vegetables for us.

Ubai's company had a tempo which we used to transport the vegetables to Mumbai. We just had to bear the diesel and driver's expenses. Neesha, who stayed in a bungalow in Bandra, offered us space for sorting and packing the vegetables. Shops like Navadanya, Green Current and Green Field, which specialized in selling organic produce, agreed to become our pickup points.

We called ourselves Mumbai Organic Farmers and Consumers Association (MOFCA) and the scheme for vegetables 'Hari Bhari Tokri Scheme'. We discussed and finalized certain norms that we would follow for the business. Our target would be the city consumers and the vegetables would have to be produced within a radius of 100 kilometres around Mumbai. This was to reduce the food miles that we would incur if we went beyond this range. The vegetables grown would be local and the seeds we gave the farmers would be open pollinated varieties. Of course the method of cultivation would be organic farming. No chemicals or pesticides would be allowed to be used.

We spoke to the farmers and agreed that they would get a fixed price for the vegetables they gave for the entire season. We would pay them the same price for all vegetables and market fluctuations would have no impact on the pricing.

We also had to ensure that the farmers strictly followed the principles of organic farming. Gaurang and I were to take the lead in this matter. We had regular weekly checks at the farms and created a form to fill in all the details of the farmer. We attended some workshops with other NGOs who were doing this kind of business and learnt how they managed to ensure quality and honesty among the farmers.

The plan was in place and we launched the scheme with great fanfare. We got some positive media reports and within a few weeks we had 200 consumers enrolled in the scheme. There were another 400 waiting to enrol. We only had a few farmers with us committed to supplying organic vegetables and so we decided to close the registration at 200 consumers.

We started the pickup from the villages and got the vegetables to Mumbai. The packing and sorting was done by volunteers who came and helped. After the initial few weeks we noticed that at the end of the pickup day, we were left with a lot of packets. This excess was becoming an issue since at that late stage we could not sell the vegetables and our losses started mounting. These vegetables had already been paid for at the farmers' end and there was no sale at the city end. We ended up either taking it home or calling friends to come and pick it up.

We sent messages to the people who did not come for the pickup and even threatened that their deposit would be lost. It was then that we realized that the deposit that we had taken was too small for the city-bred rich to even bother about. They just told us it was fine and they could not come over due to some reason or other. It was saddening to see so much wastage at the end of every pickup day.

We pulled through the sixteen weeks of delivery that we had promised and made a huge loss in the first season. The loss was mainly due to the transport cost and the wastage that had happened. But we had learnt our lesson.

The next season we changed the scheme and told all interested consumers that we would be charging for the full sixteen weeks of delivery in advance. That way we were assured that they would pick up the vegetables. We also shifted the sorting to Ubai's farm in Bhiwandi, just outside Mumbai, since he had labour that would help in sorting, packing, etc.

The encouraging news was that after seeing the success and the money earned by the four farmers of the trust, sixteen more farmers wanted to join the group and supply us. We had more vegetables and it justified the cost of transport. The more the vegetables, the less the cost of transport per kilo.

With renewed enthusiasm we started the second season. Since the number of farmers had increased and not all of them were associated with the trust, we had to come up with a stringent review system to ensure that they all followed organic principles and did not cheat by adding chemicals. This meant more regular trips to visit the farms and check the quality of the produce. We found out how the other groups were doing this review and created our own form which covered all aspects of organic farming and used it during our review visits. This way we had comprehensive data on each farmer and knew what he or she was growing and how much he or she would give us. The form had columns for even the number of saplings the farmer had planted so it gave us a good picture. Once again I was pleased that I was using all the documentation knowledge I had gathered during my years of working in the IT industry.

At the end of the second season we were better off and even had some money in surplus after paying for transport and the vegetables. It was decided that this money would be given back to the farmers as a soft loan in case they wished to buy some equipment, say a pump or some pipes, but with the clear understanding that it would be recovered from the sale of vegetables the next season.

In the third season we increased the price paid to the farmers after consulting with them. We now had twenty-five farmers from in and around the village where the trust functioned. It was extremely encouraging to see that more farmers were interested in organic farming. Besides giving them a fixed rate for the whole season to insulate them from the fluctuations of the market we also noticed that the practices we asked them to follow would in the long run increase the quality of their soil and life. We started encouraging the farmers to eat the stuff they grew organically rather than buy from the market.

On the consumer front, people started realizing what we were trying to do and were very appreciative of the whole exercise. To bring the farmers and the consumers closer we had a small get-together at the end of the second season so they could interact with the farmers directly and understand the issues. The same was true for the farmers since for the first time in their lives they were actually meeting those who consumed what they grew.

As we were making efforts to educate the farmers and try and convert them to grow organic food we realized that there was a major effort required on the consumers' end too. We were giving them local vegetables and we started getting calls from many saying they had never seen the vegetable we had sent and had no idea how to cook it. Realizing that there was a lack of awareness we started inserting a small paper in each packet given to the consumer. This insert had a recipe for the vegetable that was included on that day. Every week a new recipe was included and we got an encouraging response from the consumers who had tried out the recipes and were amazed that a vegetable which they had never seen before could taste so good.

By the end of the second season we had decided that we would have to slowly move out of the daily functions of this process since all of us had our own farms and it was difficult to invest so much time into this. We also wanted the entire activity to be self-sustaining rather than be dependent on a few committed people like us. We appointed two persons from the village itself to monitor the farmers on a daily basis. They would go to each farmer's field and see how he or she was progressing. In case they ran into some trouble they would call Gaurang or me and we would rush there and try to sort out the matter.

The sorting and packing was completely voluntary and we realized that getting volunteers on a sustained basis was not easy. Also the entire exercise was a strain on Ubai as he had to be there every week for the sorting. We tied up with a women's self-help group from a nearby village and started paying them wages to help us sort and pack. The loose informal group that we had started off with was to become more formal and after discussing various options we decided to form a company.

The work would soon no longer be dependent on volunteers but staff members of the company. We had on board all the friends who had initially thought of this idea. It was a strange feeling that after so many years of leaving the corporate world I was set to be the 'director' of a company. A full circle.

But, as time would reveal, it was not to be. I was not going to be the director in any company. We ran into a lot of legal issues in forming the company. Also, the organic market had grown since the time we started out. There were more players in the market now and they were giving the produce at a competitive price, though we still had our doubts on the organic label they attached to the produce.

We also got a lot of feedback from the consumers and there were many who did not seem happy with the basket of vegetables they got every week. We conducted a survey with our customers to find what they wanted and to also understand what was wrong with our model. The survey was extremely useful and the collation of data gave us the following results.

The consumer in Mumbai wanted:

- To choose the vegetables that they wanted
- To decide the quantity that they wanted each week
- To have the choice to refuse a supply due to various reasons

The survey also revealed that the consumer wanted to have vegetables that we could not grow in our area like potato, cabbage, cauliflower and onion.

Initially we thought we would try and keep our customers happy by procuring what they wanted from other sources. But there was the issue of transportation and also the quality and organic check. The distances from where we could procure these other vegetables would be too far for us to monitor on a regular basis and we would have to rely on their input that they really were organic. Besides, after spending so much on transport, we would be left with less money to pay the farmers which we were not comfortable with.

Finally, after a day of brainstorming, we all decided that we would have to abandon the tokri scheme and look for other alternatives to help the farmers sell their produce. It was clear that we could not leave the farmers high and dry after having worked with them for so many years to convert them to organic farmers.

We all did feel bad about terminating the tokri scheme, considering that we had put in so much effort and energy to set it up. But then when we looked at it rationally we felt that maybe it was the best we could do. We could not satisfy the Mumbai consumer by compromising on our quality and fair price to the farmer for that would have defeated the very purpose of setting up MOFCA.

We searched around and finally managed to tie up with the International Society for Krishna Consciousness (ISKCON) group who agreed to pick up the vegetables from the farms at a good price. We also found an individual who was doing home delivery in Mumbai and was willing to pick up the vegetables from the farms.

The current situation is that we have around fifty farmers in the area growing organic vegetables. We are still struggling on the marketing front as the produce is far more than what the two vendors we currently have are able to pick up. There is a need to tie up with other vendors who can pick up the excess produce. There are new players in the market but they do not promise a good price and also expect the produce to be delivered to their doorstep.

While the problems are aplenty, we have still not lost hope. We have a commitment to the farmers and are determined to find a solution to market their produce at a fair price so they continue their organic food production and ensure that the land at least is not ravaged while they try to make a living out of agriculture.

15

Are You Happy?

One of the most frequently asked questions to me is, 'Are you happy?' Many of the visitors to the farm ask me this. I could reply that if you stayed in a place like this you could not be anything but happy. Of course, it is important to understand what happiness means. It has different meanings for different people. For some people (and till recently for us too), earning lots of money, travelling trapped in a tin box with air-conditioning and an FM radio, sitting in a superb office, getting a six-figure salary, going on holidays to exotic places, owning the latest gadgets, shopping at the mall every weekend, eating at the best restaurants and watching soap operas on television every evening gives them happiness.

For us, it is different. The open sky, the beautiful scenery, our pets, the crisp vegetables, the fresh fruits and eating what we grow gives us happiness. The joy of seeing the seed you planted push out of the soil and in a few weeks turn into a huge plant is something that can never be experienced in a city mall. No Nat Geo Specials you saw would come even remotely close to seeing a buff striped keelback catch a toad and eat it in front of you, a hissing cobra just a couple of feet away or the swaying mating dance of two rat snakes or the Russell's vipers.

I do agree that we only have basic comforts and sometimes things can get a bit rough. Things that we take for granted in the city like water, electricity, communication and access to shops are all a struggle in the village. Even the lack of quality medical aid close by is worrying, especially after our encounter with the cobra and the viper. These are worrying aspects and they tend to linger in your mind, but like everything else they fade away and soon enough you are back on your feet soiling your hands with mud and carrying on with your immediate tasks.

Some people who visit me comment that I have gone back to the primitive way of living. They say this is like being a caveman. My reply to them is that we do not live like cave people. Our kitchen is well equipped, we have a gas stove, piped water, a bathroom and an erratic power connection. We have installed a battery backup which helps during power cuts though it can take the load for only a few hours. We wear good clothes, bathe regularly and keep our hair trim. We have our Internet connection and check our mails and social media sites regularly. In India, 65 per cent of the population depends on agriculture and lives in villages. Does that make us a country of cavemen?

The trick is to balance what we want to do and use gadgets and technology to the extent they are useful. I can't but mention how some of my colleagues at IBM went for a holiday in Goa with their laptops. One of them sat on the beach and worked on a proposal that we had to submit while his family frolicked in the water. I wonder what kind of a holiday he had. It's not that I have no use for gizmos or technology. The desktop at the farm is useful and it's a relief that finally network connectivity is available at the farm.

Sometimes I think all we wish to do is travel in comfort, work at a secure place and return back to our flat-screen television sets at home. We have lost out on love and sensitivity. This is because we have moved away from nature itself. By building a cocoon around ourselves, we tend to forget that it is nature which has created us. Only a flash flood or an earthquake brings us back to the reality that there exists a force unseen which needs to be respected.

A few people said that I could consider such a transition since we did not have any kids. I don't dispute this. All I can say is that this was inevitable and would have happened even if we did choose to have kids of our own. My ideas on education differ and I would prefer to have my kid trained in identifying each plant in the area than spend time imitating some silly Bollywood actor's antics on screen. Besides, right now, the farm and the pets seem like our children. We have given them all we have got and they do need tender love and care as much as a child does. What we get in return cannot be measured.

Our decision to not use chemicals is another bone of contention. Most educated and well-read people feel that these are scientific methods developed by renowned scientists. What harm would these chemicals cause anyway? There are a large number of studies which show that these harmful chemicals do enter the human system through food that has been grown using them. It is also proven that in the long run these chemicals can have a detrimental effect on health. One particular study also found the presence of urea and lead in breast milk. It is also true that these salts are retained in the soil, thereby reducing the fertility of the land and leading to loss of yield. I could see this happening in our village, a case in point being our groundnut harvest.

I am no scientist nor have I done years of research. All I can vouch for is that the food grown on our farm tastes better and lasts longer compared to what we buy from the market. The ladyfinger grown at the farm does not become sticky when cooked and the rice we grow has an aroma that no rice from the local grocer has been able to match. The idlis my mother makes from our urad dal are softer and fluffier. People who use the oil we sell call back and tell us that they like it a lot and it tastes better than commercially available ones.

The seeds we use are the same the rest of the village uses. The water we use is the same too. The only difference in their produce and ours is that they use chemicals while we do not. Nature has a way of fighting for its own. Once, one of our lemon trees was completely destroyed by the lime butterfly caterpillar. Not a single leaf was left on the plant. A few weeks later, after the lime butterfly had emerged, new leaves appeared miraculously. We got our lemon tree back and were also rewarded with a large number of lime butterflies on the farm.

There are of course some occasions when it may not be enough to sit back and watch Mother Nature. On one occasion, a vicious white grub-like insect attacked our mango tree. These tiny insects get into the trunk of the tree and slowly eat it inside out and huge trees can die in a short while. I tried a concoction of jaggery and water and poured it into the hole made by the insect. Within a few hours, friendly ants rushed into the hole after the sweet solution. It was a matter of time before they reached the insect. The ensuing battle must have been bloody and long but we could not witness it as it was deep inside the trunk. The fact was that the ants emerged victorious since the next day onwards there was no drilling by the insect.

Nature has an entire range of chemical substances available in various plants that grow around us. One of our coconut trees kept fruiting but none of the fruits survived. A bit of research revealed that this could be because of the lack of boron, a chemical substance. The option was to use boron, which is available in the market, or turn to nature for help. We read that one particular plant, calotropis, which grew wild around our village, was rich in boron. We just plucked a few stalks and put it around the coconut tree and that has made a dramatic difference.

We also realized that this kind of a setting and lifestyle does not work for everyone. One of Meena's cousins had visited us with his family. The entire family was having a roaring time at the river when we noticed that her cousin was missing. I went back to the house looking for him only to find him sitting in his car, reading the day's paper with the radio on full blast. Here was a guy who just could not adjust to the open space around him. He still needed his tiny tin space and loud music to keep him happy. There are different kinds of people in this world.

Whenever I speak to people about my experiences at the village and the new lifestyle we were leading, they would all ask me about the returns from the farm. Did you manage to 'break-even'? What is the kind of 'return on investment'? Some have even gone to the extent of trying to work out the interest I would have earned if I had put the money we invested into a financial instrument.

I have no answers to these questions. There is no money and there is no profit. All we have is a piece of land that we love and it gives us enough food to keep us both from being hungry. There is a lovely house to live in and the air is pure. We have pets we love and it's a joy living with them. The people in the village are nice and we are part of the community. I think that is more than 'break-even'. Where in the city would we be able to buy this kind of space and peace?

As for 'return on investment', the fact that we now lead healthier lives is itself a major return. Yes, I do agree that our income is low and we do have to be a little careful in our spending as compared to the IBM days when we could spend without care. Over the last few years we have realized that this did not actually affect us and we still did not miss any of the things we did earlier. It is a clear indication that all those things were not exactly essential and we could live without them. It is just a question of adjusting your lifestyle to suit your pocket.

After staying in the village for years, I also realized that you did not need much money to survive. Anyway there was hardly anything one could buy or spend on at the village. The nearest theatre is a good 40 kilometres away and there is no shopping mall around for miles. All you can buy are essentials that you need to survive. Of course, the local transport is expensive, but when at the farm we hardly travel around on a daily basis. There is an occasional expense in the form of a broken fence or a burnt-out motor but these are manageable.

Many have commented that I have taken early retirement. In reality it is far from it. The daily chores of watering, weeding and cleaning have to be done on time and they occupy a major part of your existence. Besides there is no household help and all the cooking, cleaning and washing too have to be done by you. Instead of an air-conditioned office you have to work in the hot sun soiling your feet and hands every day. At the end of the day instead of a flat-screen idiot box you have to contend with darkness and nature's theatre. Working on the farm is a full-time activity and the physical labour more than compensates for the lack of intellectual stimulation. It is only a question of time before you start enjoying the new lifestyle.

Are You Happy?

One other thing that rankles many people is the reason I wished to do this. They could not understand why a corporate executive would wish to dig his hands into cow dung and stoop to physical labour. There are many reasons for this. The chief among them is the fact that this gives me tremendous satisfaction. It is the closeness to nature that brings joy and cheer in my life. Besides I rediscovered the feeling of living in a community at the village. It brought back pleasant memories of my childhood and my years at the railway colony.

I do agree that people migrate to the city due to poverty. Many really don't have a choice. But in a sense I was convinced that for some it was not life and death but the need for more money, better jobs and more materialistic comforts that pushed them into the city. They are prepared to undergo tremendous hardship and live in abysmal conditions in the city. However, people who lived in the city have settled down very well in villages and I was inspired and guided by many friends who had already done this. It is also true that at the village and farm I found peace and tranquillity. It is evident from my well-being itself that this kind of lifestyle has had a positive effect on me.

I read somewhere that the wellness of each individual is based on five parameters. They are the financial, the emotional, the social, the physical and the spiritual. When all the parameters mentioned are in balance one can say that one is completely 'well'. Currently, in the city, the most important is the financial parameter. It is presumed that the rest will automatically follow. It is unfortunately true that money can never buy you friends or health or mental peace.

Over a period of time, after staying in the village, I realized that this financial aspect of life was slowly but surely creeping into the villages too. Here too people are looking only at money and ways of earning it. I could see boys like Hemant moving away from farming and hoping to make millions. Why was this happening? Was it because of the flawed education that we were imparting to these youngsters? Was it the failing yields from the land that made them turn away from agriculture? Was it the media portrayal of farming as a poor man's occupation? I am still not able to arrive at a single answer to this question.

Sometimes, I felt it was the invention of money itself that was responsible. If money did not exist we would not have this problem at all. Our villages are more than self-sufficient in their needs. What they do not have they can barter from other villages which have what they need. I saw that there still existed a semblance of the barter system in our village too, especially at the Mahalaxmi temple fair.

Almost every other day we read reports of farmers committing suicide all over the country. It is not surprising, considering that they pump in so much money to buy better seeds and more expensive chemicals to protect their crops. Besides, after the effort and toil they put in, the returns do not even cover the cost of production, let alone give them a decent profit. Since they do not have any savings or ready cash they borrow from unscrupulous sources who charge high interest rates and cheat them. The only solution I can think of is to reduce the input costs which can be done by turning to organic agriculture and of course by getting fair prices with some assistance from the government.

It is not that the government has no schemes or is not trying to help the farmers. There are numerous schemes under various heads that are applicable to all farmers. Unfortunately, these schemes never reach the average farmer and are used only by a select few who get to know of these. The information on these schemes is supposed to be given to the villagers by either the gram sevak or the sarpanch of the village. It never reaches the villagers and only those close to the sarpanch or those clever enough to find out for themselves benefit. Most of these schemes also need an initial small investment which only the rich or at least well-to-do farmers can afford. Besides most of them opt for the scheme to get the subsidy money and are not really interested in the output of the schemes.

In the quest for money from other sources, farmers send their children to far-off factories to work as unskilled labour. These factories exploit them and pay them far below the minimum wages prescribed by the government. Yet, there is no other option. Many are indebted and desperately need an income just to survive. It is one of the main reasons for the large-scale migration happening to the metro cities. This in turn means that the agriculture in the village is done mostly by older men and women while the younger generation is in far-off places to earn money.

The other problem due to this large-scale migration is the loss of traditional knowledge at the village level. Most of the agricultural know-how and intelligence is passed on from one generation to the other verbally. With the younger generation opting out of farming, this priceless traditional knowledge is getting lost. As I have mentioned earlier there are no manuals or notes in this occupation.

It is rather unfortunate that the government, in its zeal to bring about 'development' in the rural areas, is in fact actually ruining their way of life. For years the villagers have been using barks and leaves as soap to wash themselves and their clothes. In the name of modern lifestyle, they have been introduced to detergents and bar soaps. Unfortunately, this is not available naturally and has to be bought from the market, requiring the all-important cash. The common beedi is being pushed out by the filter cigarette while the bamboo basket is being edged out by plastic bags. The common sea salt has been edged out by the packaged iodized salt which is far too expensive than the former. What the villagers got locally is now being imported from far-flung factories and to buy these things they need money.

To add to this situation, the media and society portray people who use traditional and natural products as some sort of primitive cave people while the new modern products are promoted by stars and celebrities. This media onslaught has its impact and it is now not a surprise when a villager offers you soft drinks and packaged chips.

What's in the Future?

I have been around at the farm for fourteen years now, much to the surprise of the villagers. They had expected me to return to the city in a year or two. Even sceptics like Moru Dada who had helped me get this land realized later that I was not abandoning my efforts and going back. The original plan we had on paper is still valid and on track. We have still not dug into our capital which grew after Meena's stint with *The Hindu* and the farm is running from my savings and some money from the groundnut oil, chikoos, mangoes and other crops. There are a few things in the plan that have not yet been executed. I still have to start keeping a few cattle.

Farming is not demeaning nor is it unremunerative. It is just that it does not generate the huge incomes that most people feel is needed to live a satisfied life. It is possible for a family to survive just by farming and tilling their land. They can get what they need for their daily existence from the land and sell off the surplus, if any. Land degradation is a major issue as is landlessness. The barter system prevalent in many parts is a way out of dependence on money.

The zeal to generate more cash pushes the farmers into the vicious debt cycle. Loan availability to farmers is a joke and this, coupled with the globalization process where world prices dictate terms, spells ruin for the farmer. Indian farmers do not have the huge subsidies that European or US farmers enjoy. The lure of cash crops is strong and they tend to forget the traditional produce that would at least feed their families and keep them alive. True wealth is in being happy, healthy and content with what one has.

I cannot bring myself to think of what will happen to the agricultural land that the next generation will own. Most of the young generation are working in companies or studying in schools and have no intention of farming at all. They are already migrating to the city to find a job and settle there. They will be part of the GenNext of this country. If this is the situation in all villages I cannot imagine what will happen to our food security in a few years. We will probably have to start importing canned stuff from abroad as fresh food would have vanished from this country. Or will we let contract farming take over?

Is this where we are headed? A scary thought.

Glossary

Ashram Shala	A residential school for Adivasi children run by the government
Basti	A cluster or settlement of houses
Bhakri	Round, flat unleavened bread usually made of rice or millets
Chapra	A small hut or sit-out
Gilli danda	An amateur sport played using two sticks; the large one, danda, is used to hit the small one, gilli
Gram	Village
Gram scvak	A government employee appointed to advise and assist villagers in matters of community welfare and development
Janmashtami	The annual Hindu festival celebrating the birth of Lord Krishna
Jatra	A fair
Kunbis	A caste of traditional non-elite tillers now included in Other Backward Classes
Mahamandal	Corporation
Pav	A soft bread roll
Sepoy	Earlier meaning 'infantry soldier', now used to refer to a doorman or peon
Warlis	Indigenous tribe living in the coastal areas of Maharashtra and Gujarat